Springer-Verlag France S.A.R.L

André Leblanc

ATLAS
OF HEARING AND BALANCE
ORGANS

A Practical Guide for Otolaryngologists

Forewords by J. P. Francke, Y. Guerrier, and C. Frèche

Springer

André Leblanc
Service de Radiologie
Centre Hospitalier Régional et Universitaire d'Amiens
80030 Amiens, France

Personal address:
20, rue Sainte Colombe
80800 Aubigny, France
Fax: 33 3 22 48 31 23

Library of Congress Cataloging–in–Publication Data
Leblanc, Andre, 1934
[Atlas des organes de l'audition et de l'équilibration. English]
Atlas of hearing and balance organs / Andre Leblanc ; forewords by J. P. Francke, Y.
Guerrier, and C. Freche ; translated by lucia Huffman–touzet.
P. cm.
Includes bibliographical references and index.
ISBN 978-2-287-59648-3   ISBN 978-2-8178-0775-1 (eBook)
DOI 10.1007/978-2-8178-0775-1
1. Ear -- Atlases. 2. Labyrinth (Ear)--Atlases. 3. Ear--Magnetic resonance imaging--
Atlases. 4. Ear--Tomography--Atlases.
I. Title.
[DNLM: 1. Ear--anatomy & histology atlases. WV 17L445a 1998a]
QM507.L3813 1998
611' .85' 0222--dc21
DNML/DLC
for Library of congress

Atlas of Hearing and Balance Organs

ISBN 978-2-287-59648-3

© Springer-Verlag France 1999
Originally published by Springer-Verlag France in 1999

Cover illustration: André Leblanc

English translation by Licia HUFFMAN - TOUZET, 16, rue Ober 59118 Wambrechies - France
(Except for pages 8, 14, 42, 43, 48, and 50)

SPIN: 10682799

Foreword

Twenty years ago André Leblanc first walked into the department to present his manuscript on the determination of the axes of the various foramens, canals, and sulci of the base of the skull, their tomographical investigation and their environment, to my master, Professor Claude Libersa. The project was thoroughly remodeled and enhanced by coupling classical anatomy with the exploding new imaging techniques.

Only a curious, minute, inventive, and tireless worker like André Leblanc could make this ambitions project a success. Thanks to his determination, he mobilized some of the best radiologists, clinicians, editors, and even anatomists, and urged each one of them on toward excellency. I admit that at times, we felt annoyed but we have forgiven him for the sake of his rigourous demonstration and admirable results.

André Leblanc's volume entitled "The Cranial Nerves", which was first published in French in 1989, and later in English, quickly became the definitive reference book for all those who deal with the cranial nerves, whether on a regular basis or occasionally. Thus a new updated edition was published both in French and English in 1995.

While others would have savoured their success, André Leblanc never stopped working, running from one congress to the next, and charming everybody from Chicago to Singapore to Taiwan... Not a month goes by without one of André Leblanc's new posters, more educational than ever, being added to the others to the walls of radiology practices or MRI centers.

The book we present today, entitled "Atlas of Hearing and Balance Organs A Practical Guide For Otolaryngologists", is a model of its kind in terms of rigour, knowledge, and aesthetics. The new perspectives that it offers will help each and everyone get a better grasp of the ear's organization and identify the 70 elements that are said to compose it.

We extend our deepest thanks to André Leblanc, for he is a wonderful teacher, and an heir, as Pierre Lasjaunias once said, to the French clinical and anatomical tradition. His many publications open a remarkable window onto the fields of anatomy and imaging. As Yves Guerrier stated without hesitation, here is a scientist whose reputation has reached far beyond our borders.

Professor Jean Paul Francke
Department of Anatomy and Organogenesis
University Lille 2 "Law and Health Care" (France)

Foreword

It is an honour and a pleasure to write a foreword to Mr. André Leblanc's outstanding publication.

I have known him for many years. His exceptional talent has already been recognized, and will probably be appreciated worldwide in the forthcoming months.

It is a great honor for the French Society of Otolaryngology, of which I am president, to have contributed to such a success.

Professor Charles Frèche
Secretary General,
French Society of Otolaryngology
and Maxillofacial Pathologies
Head, Department of Otolaryngology
American Hospital of Paris, Paris

Foreword

André Leblanc is an extremely talented illustrator of anatomy. He ranks among the most famous graphists of morphological textbooks published in the last few years.

He has now chosen to depict the auditory and vestibular pathways: nothing can be more abstract, as are the perilymphatic spaces that he represented with talent and precision. For those of us who already know, and even more so for those who are learning, these structures are very elaborate, and have been called, rightly so, the labyrinth.

Those who should know the constitution of the ear often didn't take time to look into it because, they said, it was too complex and too tedious. They will no longer be able to say that: the beauty and the precision of André Leblanc's drawings make the study of anatomy pleasant.

I would have liked to work with him, much as an author likes to write for a gifted musician.

Professor Yves Guerrier
President, French Society of
Cervicofacial Oncology;
National Corresponding Member of
the Academy of Medicine
75, avenue de Lodève, Montpellier (France)

Preface and Acknowledgements

This volume on the internal ear adds substantial information to the eighth chapter of the second edition of my atlas entitled "The Cranial Nerves" (1995). Many preliminary explorations on dry bone and dissections were necessary to develop it. In contrast, imaging in a multitude of planes seems effortless, thanks to the angles that I have defined and tried on many morphologically very different patients over the years. These reference angles enable the investigator to determine the precise axis of an orifice or to view the course of a nerve. They can be used regardless of the morphology or the state of the patient, and take possible asymmetries into account.

This new publication entitled "Atlas of Hearing and Balance Organs" is a simple yet original approach to anatomical investigation and imaging. When used as a guide (the auditory tract is depicted by means of serial macroscopic sections and dissections, but also imaging), most of the vulnerable parts of the nerve and the vestibulocochlear pathway can be quickly visualized.

The benefits of this book lie not only in the combination of anatomy with modern imaging techniques (CT and MRI), but also and predominantly in the numerous diagrams of bony fenestrations of the cochlea, the vestibule and the semicircular canals. These views reveal the membranous labyrinth, the internal organs of balance and audition, and highlight their innervation, as well as the utricular and saccular nerves, the nerve of the spiral organ of Corti....

The vestibulocochlear nerve is depicted from its true (nuclei) and apparent (bulbopontine sulcus) origins, on through its course in the canal, and to the innermost part of the ear. The endolymphatic system is described by means of computed tomographies and shows the aqueduct of the vestibule and the endolymphatic duct. The study of perilymphatic space also relies on computed tomography to evidence the aqueduct of the cochlea and the perilymphatic duct.

Vasculature of the auditory tract is traced from the vertebral artery to the arteries of the internal ear.

Constant progress in new imaging techniques has broadened the possibilities of oblique planes and three-dimensional reconstruction. This new "method" can thus be easily adapted to new technologies, although the reading of the views may prove more delicate.

It is a fact that clinicians and radiologists need extremely detailed anatomical references. This work will be a precious guide for them in the areas of anatomy and imaging techniques.

Let us list a few of the most relevant aspects of the book:
- a guide for otolaryngologists, neurologists, anatomists, and radiologists;
- valuable teaching material for this difficult-to-explore area;
- assistance the investigation of temporal neuralgia and otalgia, early diagnosis of neurinoma, otosclerosis, cholesteatoma, and tumoral formations, though limited in size, from the onset of clinical signs;
- easier exploration, regardless of the state and morphology of the patient;
- a means to obtain CT and MRI views in minimum time thanks to the precisely defined reference angles;
- a channel to renew interest in this part of the body also due to the progress in tomography and in magnetic resonance imaging. The quality of spatial resolution now makes it possible to visualize the meanders in the canal, the intracranial course of the nerves, and the cavities of the inner ear.

This book is the fruit of 40 years of work and research, during which I was assisted by the Institute of Anatomy of the Faculty of Medicine in Lille, and in particular by Professor J. P. Francke, whom I wish to thank for the anatomical sections and dissections that he has faithfully carried out since 1980 every time a chapter required this type of illustration.

I wish to dedicate this book to Professor Claude Libersa. By his presence and expertise, he gave me the means to pursue my study of an arduous discipline when others didn't attribute much credit to this work.

Thanks to his generous support and his determination, the many posters illustrating the course and vasculature of the cranial nerves, as well as the first two editions of the atlas dedicated to the cranial nerves, were able to be completed and published, and thus satisfy the needs of medicine and modern science.

Today the third edition of the atlas, which is under way, and this volume more specifically geared toward otolaryngologists, came to be because of him, and I wish to express my gratefulness to him.

Contributors

Jan W. Casselman
Doctor
Neuroradiology and Radiology of Head and Neck
A.Z. St Jan, Brugge
Ruddershove (Belgium)

Chun Siang Chen
Research Assistant Professor
Department of Neurological Surgery
Mount Sinai Medical Center
New York, NY (USA)

Jean Paul Francke
Professor
Department of Anatomy and Organogenesis
University of Lille 2, Lille (France)

Charles Frèche
Professor
Secretary for the French Society of Otolaryngology
and Cervicofacial Disorders
Department of Otolaryngology
American Hospital of Paris, Paris (France)

Yves Guerrier
Professor
Department of Otolaryngology and Cervicofacial Disorders
Member of the National Academy of Medicine
Montpellier (France)

Claude Libersa
Professor
Department of Anatomy and Organogenesis
University of Lille 2, Lille (France)

Jean-Claude Libersa
Professor of Dental Pathology and Therapeutics
University of Lille 2, Lille (France)

Gabriele Meier
Editor
Nibelungenstr. 55
München (Germany)

Kalmon D. Post
Professor and Chairman
Department of Neurological Surgery
Mount Sinai Medical Center
New York, NY (USA)

Chandranath Sen
Associate Professor and Vice Chairman
Department of Neurological Surgery
Mount Sinai Medical Center
New York, NY (USA)

Vladimir Strunski
Professor
Department of Otolaryngology
University Hospital of Amiens
Amiens (France)

Important Notice

Before reaching the muscles they supply, the cranial nerves pass through a number of foramina, canals, and sulci. Each one of these orifices has a very specific axis, and it is along this axis that it should be viewed.

Traumatism on the skull can cause a number of pathologies ranging from anaesthesia to paralysis. They are the result of either haematomas due to a fracture, or tumoral lesions which compress the canal containing the nerve.

Each structure through which the cranial nerves pass has its vulnerable points. Therefore great care should be taken to explore each channel along its own axis, for fear of deforming the image and thus, making the wrong diagnosis.

A few years back imaging relied exclusively upon conventional X-ray machines, which mainly provided views of the bones. When exploring a cranial nerve related pathology, radiological examination was limited to viewing the bony orifice of interest, exactly along its axis, based on two reference lines. These reference lines were constant, regardless of the morphology and the state of the patient, and took possible asymmetries into account.

Thus, conventional X-ray investigations required a very detailed knowledge of anatomy. Unfortunately, modern imaging techniques have contributed to the decline of anatomy as a discipline.

It is true that computed tomography and magnetic resonance imaging have become the main tools in intracranial exploration, for they provide the likes of an anatomical section. Nevertheless, CT and MRI views are rarely taken along the axis of the ostia and canals. Although the images are three-dimensional, and because most channels are extremely sinuous, the operator can fail to notice a small fracture, a haematoma, or a minute lesion compressing a nerve if proper orientation of the views is not respected.

Reliable views can be obtained by following the conventional technique based on centreing and anatomical references. This is also possible with CT and MRI, thanks to the broad possibilities of orientation and three-dimensional reconstruction.

Thus the conventional method can easily be adapted to modern imaging. Interpretation is rendered more delicate, it is true. But clinicians and radiologists must refer to detailed anatomical information.

That is why the author continues to present the descriptions and diagrams, which he has adapted to modern imaging techniques, of conventional reference angles in both the present volume and the third edition of his atlas entitled "The Cranial Nerves" © 2000.

André Leblanc

Contents

Foreword (Professor J. P. Francke) V

Foreword (Professor C. Frèche) VI

Foreword (Professor Y. Guerrier) VII

Preface and Acknowledgements VIII

Contributors IX

Important Notice X

Organs of Hearing and Balance 1

References 51

Index 53

Presentation of Posters 56

ANATOMY (COURSE–DISTRIBUTION–BRANCHES)
Topography of imaging correlated to anatomy. . . 2
Topographical description 3

**TRUE ORIGIN OF THE VESTIBULOCOCHLEAR
NERVE**
VESTIBULAR AND COCHLEAR NUCLEI
Anatomy, diagrams, and MRI views 4, 5

**APPARENT ORIGIN OF THE VESTIBULO-
COCHLEAR NERVE**
BULBOPONTINE SULCUS
Anatomy, diagrams, and MRI views 6–9

VESTIBULAR AND COCHLEAR NERVES
CEREBELLOPONTINE ANGLE – EXTERNAL
ACOUSTIC MEATUS
Anatomy, and MRI views 10, 11

**INTERNAL EAR AND VESTIBULOCOCHLEAR
PATHWAYS**
ORGANS OF HEARING AND BALANCE
SPIRAL ORGAN OF CORTI, AMPULLARY CRESTS,
MEMBRANOUS LABYRINTH,
MODIOLUS, UTRICLE, SACCULE, VESTIBULE,
SEMICIRCULAR CANALS
Anatomy, diagrams, CT and MRI views . . . 12–28

ENDOLYMPHATIC SYSTEM
ENDOLYMPHATIC SAC, ENDOLYMPHATIC DUCT,
VESTIBULAR AQUEDUCT
Anatomy, diagrams and CT views 29, 30

PERILYMPHATIC SPACE
PERILYMPHATIC DUCT
COCHLEAR AQUEDUCT
Anatomy, diagrams and CT views 29–31

LABYRINTHINE VASCULATURE
ANTERIOR INFERIOR CEREBELLAR ARTERY,
LABYRINTHINE ARTERY, VESTIBULAR AND
COCHLEAR ARTERIES
Anatomy, diagrams and MRI views 32–40

MIDDLE EAR, OSSICULAR CHAIN
INCUDOMALLEAR AND INCUDOSTAPEDIAL
ARTICULATIONS
CAVITY OF THE MIDDLE EAR, EPITYMPANIC
RECESS
Anatomy, diagrams and CT views 41–47

AUDITORY TUBE
CHANNEL OF THE AUDITORY TUBE, TYMPANIC
AND PHARYNGEAL ORIFICES
OF THE AUDITORY TUBE, PHARYNGEAL
(or ROSENMÜLLER'S) RECESS, CAVUM
Anatomy, diagrams and CT views 48, 49

EXTERNAL EAR
EXTERNAL ACOUSTIC MEATUS
Anatomy, diagrams and CT views 50

REFERENCES 51

INDEX 53

REDUCED VERSION OF SIX POSTERS ON THE EAR
1. Organs of hearing and balance
2. Cavities of the inner ear
3. Labyrinthine vasculature
4. Vasculature and cavities of the
   organs of hearing and balance
5. Vestibulocochlear pathways
6. The internal ear 56–58

Anatomy

ORIGIN – DISTRIBUTION – COLLATERALS

To study these structures

Real and apparent origins, intracisternal course.

Intracanalar course.

Spiral organ of Corti, cochlear ganglion.
Vestibular ganglion, utricle, saccule, ampullary crests.

Endolymphatic system.
Perilymphatic space.

Ossicles, middle ear.

Auditory tube, cavities of the middle ear.

Imaging

AREAS EXPLORED

Explore the areas

Pontine cistern, study of the bulbopontine sulcus and the inferior portion of the pons for the anterior part of the fourth ventricle.

Investigation of the external acoustic meatus.

Study of the cochlea, vestibule, and semicircular canals.

Exploration of the semicircular canals, vestibule and cochlea to visualize the aqueduct of the vestibule and the endolymphatic duct for the endolymphatic system, and the aqueduct of the cochlea for the perilymphatic space.

Views of the incudomallear and incudostapedial articulations (hearing ossicles) and of the epitympanic recess.

Exploration of the auditory tube, the pharyngeal recess, the epitympanic recess and the mastoid antrum.

The vestibulocochlear nerve is a sensory nerve. It consists of two parts: the cochlear nerve and the vestibular nerve. The nerve enters the pons at the lateral extremity of the medullopontine sulcus, lateral to the facial nerve and a little above and in front of the glossopharyngeal nerve.

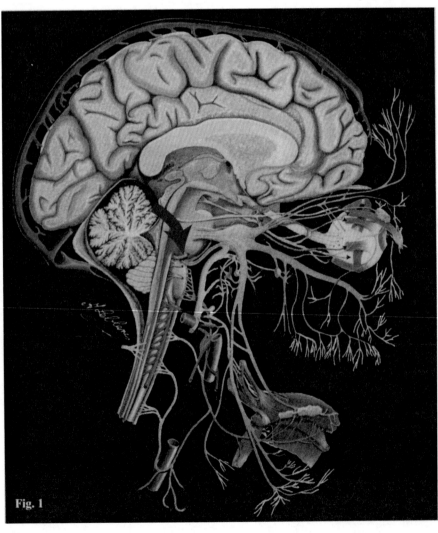

Fig. 1

Topographical description

Real origins of the vestibular and cochlear nerves (nuclei)
– Inferior pons
– Floor of the fourth ventricle

Apparent origin of the vestibulocochlear nerve
– Bulbopontine sulcus
– Pontine cistern

Vestibular nerve, vestibular ganglion
Utricle, saccule, ampullary crests
– Internal acoustic meatus
– Vestibule
– Semicircular canals

Cochlear nerve, cochlear ganglion
Spiral organ of Corti
– Internal acoustic meatus
– Cochlea

Endolymphatic system
– Vestibule
– Semicircular canals
– Cochlea
– Cochlear duct
– Membranous labyrinth
– Aqueduct of vestibule
– Endolymphatic duct

Perilymphatic space
– Vestibule
– Semicircular canals
– Cochlea
– Scalae vestibuli and tympani
– External orifice of perilymphatic duct
  (petrosal fossula of IX)
  and perilymphatic duct

Internal ear

Middle ear
– Tympanic cavity
– Ossicular chain
– Epitympanic recesses
– Mastoid antrum
– Epitympanic space

Nasotubal orifices of the middle ear
– Auditory tube
– Pharyngeal or Rosenmüller's recess
– Nasal cavity
– Epitympanic recess
– Mastoid antrum

External ear
– External acoustic meatus

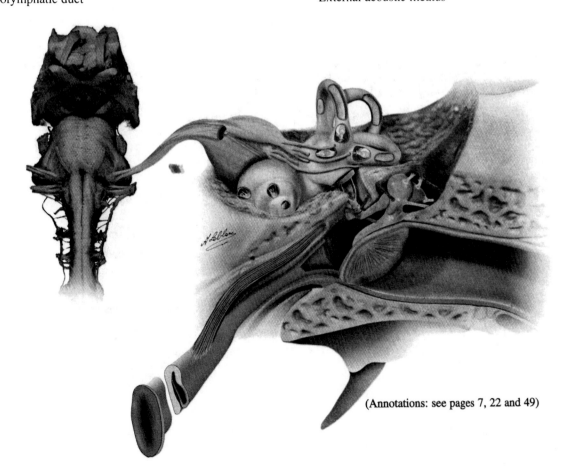

(Annotations: see pages 7, 22 and 49)

Fig. 2 Diagram of internal, middle, external ear and organs of hearing and balance

TRUE ORIGIN OF THE VESTIBULOCOCHLEAR NERVE
VESTIBULAR AND COCHLEAR NUCLEI
Anatomy, diagrams, and MRI views

Cochlear and vestibular nuclei

The **cochlear nuclei** collect information from the inner and outer hair cells contained in Corti's spiral organ via the cochlear root of the vestibulocochlear nerve. Dendrites pass through the spiral lamina and reach the spiral canal of the modiolus (Rosenthal's canal) (Fig. 12. e; 15. b, h).

The cochlear nuclear complex is located on the dorsolateral side of the brain stem, next to the bulbopontine junction. It consists in a ventral nucleus and a dorsal nucleus. The anterior nucleus rests on the anterior and exterior face of the inferior cerebellar peduncle, while the posterior nucleus is continuous with the lateral recess of the fourth ventricle (Fig. 3. a– f).
This is where an electrode should be placed for central stimulation.
The secondary sensory neurons in the cochlear path extend to the medial geniculate body after 80% of them have decussated. They form a polysynaptic chain called the lateral lemniscus.
The terminal neurons arise from the geniculate body and end in the transverse temporal gyrus (area 41).

The **vestibular nuclei** collect the impulses from the base of the ciliated cells in the vestibular epithelium that coats the labyrinth, e.g., the semicircular canals, utricle, and saccule (Fig. 14. b– g; 15. e, f).
Dendrites belonging to the vestibular nerve form two roots: the superior vestibular root, resulting from the merging of the utricular nerve and fibers from the ampullary crests on the anterior and lateral semicircular canals; and the inferior vestibular root, which is constituted by the fibres from the saccular macula and the posterior semicanal. Both roots reach the vestibular ganglion located in the most posterior part of the external acoustic meatus (Fig. 4; 11. b, c; 14. c; 15. f).
Classically, the vestibular nuclei, located at the bulbopontine junction, are divided into four groups: the superior vestibular nucleus (Bekhterev's nucleus), the medial, lateral, and inferior vestibular nuclei. They connect with the archeocerebellum, the thalamus, and the cerebral cortex via the central vestibular pathway, and with the spinal cord via the vestibulospinal tract (Fig. 3. a–f).

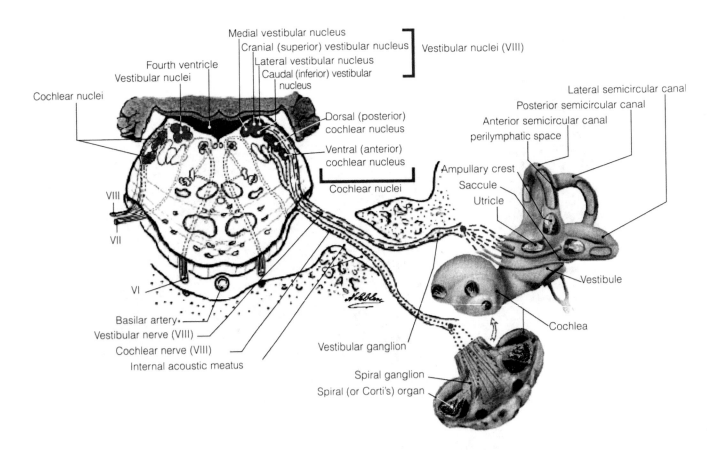

Fig. 3 a–f. Diagram of vestibulocochlear nuclei (**a**), red dotted lines: cochlear nuclei; green dotted lines: vestibular nuclei; axial MRI view at the level of the pons for the nuclei (**b–e**); posterior anatomical view of the brain stem at the level of the floor of the fourth ventricle showing the position of the vestibulocochlear nuclei (**f**) (MRI: Dr. J. W. Casselman, A.Z. St Jan, Brugge)

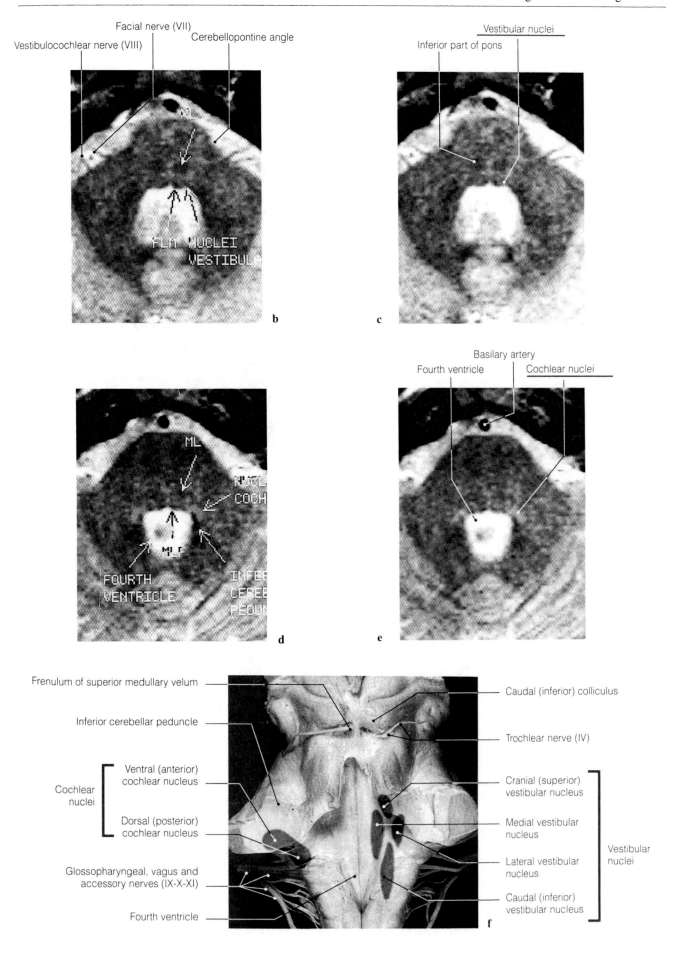

b

Vestibulocochlear nerve (VIII)
Facial nerve (VII)
Cerebellopontine angle

ML

FLM NUCLEI
VESTIBUL

c

Inferior part of pons
Vestibular nuclei

d

ML

NUCL
COCH

MLF

FOURTH
VENTRICLE

INFER
CEREB
PEDUN

e

Fourth ventricle
Basilary artery
Cochlear nuclei

f

Frenulum of superior medullary velum

Inferior cerebellar peduncle

Cochlear nuclei
    Ventral (anterior) cochlear nucleus
    Dorsal (posterior) cochlear nucleus

Glossopharyngeal, vagus and accessory nerves (IX-X-XI)

Fourth ventricle

Caudal (inferior) colliculus

Trochlear nerve (IV)

Cranial (superior) vestibular nucleus

Medial vestibular nucleus

Lateral vestibular nucleus

Caudal (inferior) vestibular nucleus

Vestibular nuclei

APPARENT ORIGIN OF THE VESTIBULOCOCHLEAR NERVE
BULBOPONTINE SULCUS
Anatomy, diagrams, and MRI views
(Pages 6–9)

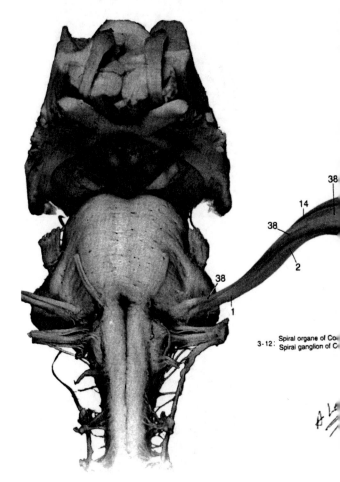

Fig. 4

1 Vestibulocochlear nerve (VIII)	**VESTIBULAR NERVE (VIII)**	24 Lateral ampullary nerve
	14 Vestibular nerve (VIII)	25 Ampullary sulcus
COCHLEAR NERVE (VIII)	15 Superior vestibular ganglion of Scarpa	26 Ampullary crest
2 Cochlear nerve (VIII)	16 Anterior ampullary nerve (VIII)	27 Transitional zone
3 Spiral ganglion of Corti (VIII)	17 Anterior ampullary nerve entering ampullary crest	28 Lateral semicircular canal
4 Scala tympani	18 Neuroepithelium	29 Posterior ampullary nerve
5 Outer hair cells	19 Hair cells	30 Posterior semicircular canal
6 Cochlear duct	20 Sensory hair of cupula	31 Utricular nerve (VIII) entering utricle
7 Scala vestibuli	21 Osseous labyrinth	32 Stereocilia
8 Inner sulcus cells	22 Membranous labyrinth	33 Outer hair cells
9 Hensen's cells	22b Perilymphatic space	34 Membranous wall of utricle
10 Stereocilia	23 Anterior semicircular canal	35 Saccular nerve
11 Tectorial membrane	23b Common bony canal	36 Vestibule
12 Vestibular membrane		37 Nerve of singular foramen
13 Cochlea		

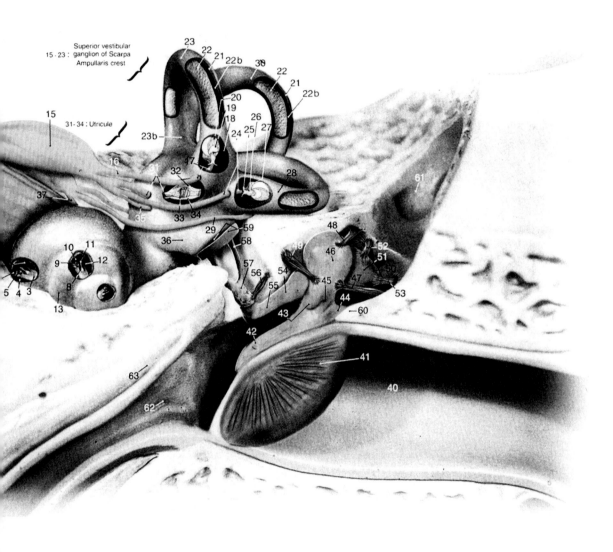

Superior vestibular
15 - 23 : ganglion of Scarpa
Ampullaris crest

15

31 - 34 : Utricule

23b

FACIAL NERVE (VII)

38   Facial nerve (VII)
39   Intermediate nerve
      [Wrisberg] (VIIb)

MIDDLE AND EXTERNAL EAR

40   External acoustic meatus
41   Tympanic membrane
42   Handle of malleus
43   Anterior process of malleus

44   Lateral process of malleus
45   Neck of malleus
46   Head of malleus
47   Lateral ligament of malleus
48   Superior ligament of malleus
49   Malleus muscle
50   Short crus of incus
51   Body of incus
52   Superior ligament of incus
53   Posterior ligament of incus
54   Long crus of incus

55   Lenticular process of incus
56   Stapedius muscle
57   Head of stapes
58   Posterior crus of stapes
59   Oval window and base of stapes
60   Superior epitympanic recess
61   Mastoid antrum
62   Auditory tube
63   Tensor muscle of tympanum

Anatomy

The **vestibulocochlear nerve** is a sensory nerve and consists of two parts: the cochlear nerve (for audition) and the vestibular nerve (for balance).

The **cochlear nerve** joins the vestibular nerve, traverses the internal acoustic meatus, becomes intracisternal and then enters the neuraxis via the lateral part of the medullopontine sulcus. It terminates in the cochlear nuclei at the lower part of the pons: the anterior nucleus and the dorsal nucleus.

The **vestibular nerve** gathers the auditory impressions in the internal ear and then transmits these centrally.

POSSIBLE CAUSES: Inferior protuberantial syndrome (Foville's syndrome), acousticofacial neurinoma (VII - VIIb - VIII) at the level of the cerebellopontine angle, lesions of the pons and the medulla oblongata.

EXAMINATION: Study by magnetic resonance imaging (MRI) or computed tomography (CT) of the cerebellopontine angle of the medullopontine sulcus, so as to visualize the apparent origin of the vestibulocochlear nerve (VIII) in sagittal and axial views and the Worms-Bretton semiaxial view.

The sections in the Worms-Bretton view are made starting from the level of the sella turcica up to below the external acoustic meatus. The technique of examination is identical to that used for the apparent origin of the facial nerve (VII).

Note: In order to eliminate any possibility of an intracanalicular lesion (ballooning of the internal acoustic meatus), it is necessary to make a comparative study of the internal acoustic meatuses should be carried out using the intraorbital survey view before embarking on this study.

Cerebellopontine angle
Magnetic resonance imaging (MRI)
AXIAL VIEWS

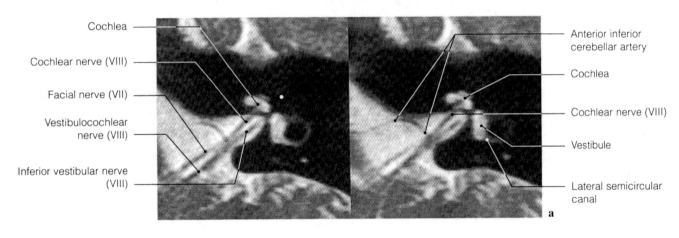

Cochlea — Anterior inferior cerebellar artery — Cochlea — Cochlear nerve (VIII) — Vestibule — Lateral semicircular canal — Cochlea — Cochlear nerve (VIII) — Facial nerve (VII) — Vestibulocochlear nerve (VIII) — Inferior vestibular nerve (VIII)

a

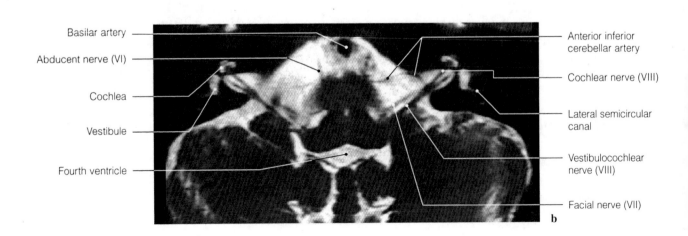

Basilar artery — Abducent nerve (VI) — Cochlea — Vestibule — Fourth ventricle — Anterior inferior cerebellar artery — Cochlear nerve (VIII) — Lateral semicircular canal — Vestibulocochlear nerve (VIII) — Facial nerve (VII)

b

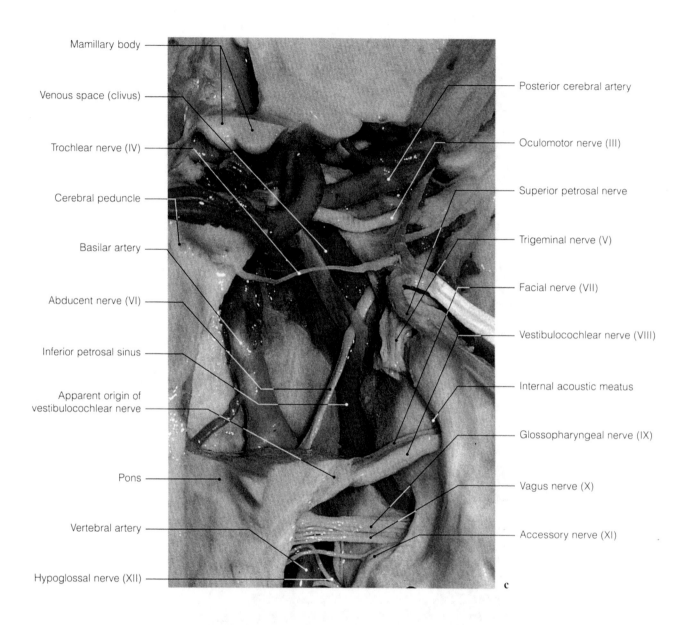

Mamillary body

Venous space (clivus)

Trochlear nerve (IV)

Cerebral peduncle

Basilar artery

Abducent nerve (VI)

Inferior petrosal sinus

Apparent origin of vestibulocochlear nerve

Pons

Vertebral artery

Hypoglossal nerve (XII)

Posterior cerebral artery

Oculomotor nerve (III)

Superior petrosal nerve

Trigeminal nerve (V)

Facial nerve (VII)

Vestibulocochlear nerve (VIII)

Internal acoustic meatus

Glossopharyngeal nerve (IX)

Vagus nerve (X)

Accessory nerve (XI)

c

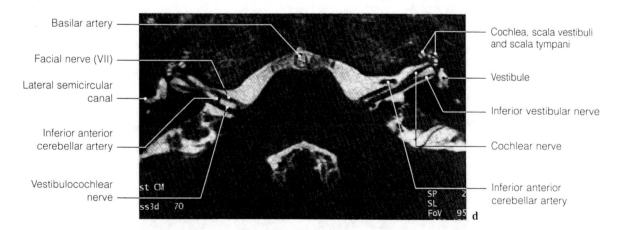

Basilar artery

Facial nerve (VII)

Lateral semicircular canal

Inferior anterior cerebellar artery

Vestibulocochlear nerve

Cochlea, scala vestibuli and scala tympani

Vestibule

Inferior vestibular nerve

Cochlear nerve

Inferior anterior cerebellar artery

st CM
ss3d   70

SP   2
SL
FoV 95   d

Fig. 5 a–d MRI views (**a, b, d**) and posterior anatomical view (**c**) of the brain stem showing the apparent origin of the vestibular and cochlear nerves (anatomical view: Prof. C. Sen, Prof. C. S. Chen, Prof. K. D. Post, *Microsurgical Anatomy of the Skull Base*, Thieme, 1997)

VESTIBULAR AND COCHLEAR NERVES
CEREBELLOPONTINE ANGLE – EXTERNAL ACOUSTIC MEATUS

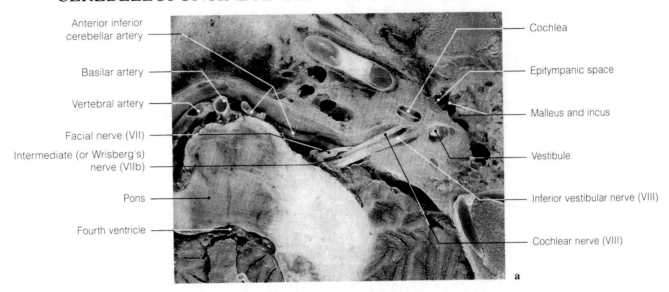

Anterior inferior cerebellar artery —
Basilar artery —
Vertebral artery —
Facial nerve (VII) —
Intermediate (or Wrisberg's) nerve (VIIb) —
Pons —
Fourth ventricle —

— Cochlea
— Epitympanic space
— Malleus and incus
— Vestibule
— Inferior vestibular nerve (VIII)
— Cochlear nerve (VIII)

a

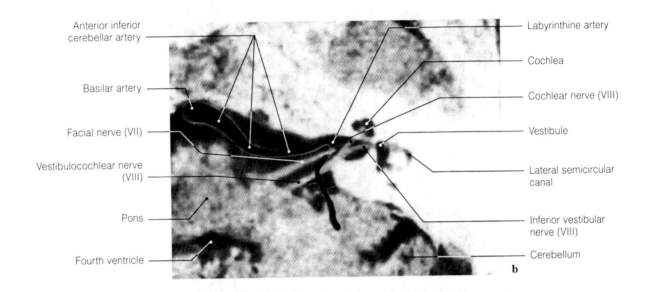

Anterior inferior cerebellar artery —
Basilar artery —
Facial nerve (VII) —
Vestibulocochlear nerve (VIII) —
Pons —
Fourth ventricle —

— Labyrinthine artery
— Cochlea
— Cochlear nerve (VIII)
— Vestibule
— Lateral semicircular canal
— Inferior vestibular nerve (VIII)
— Cerebellum

b

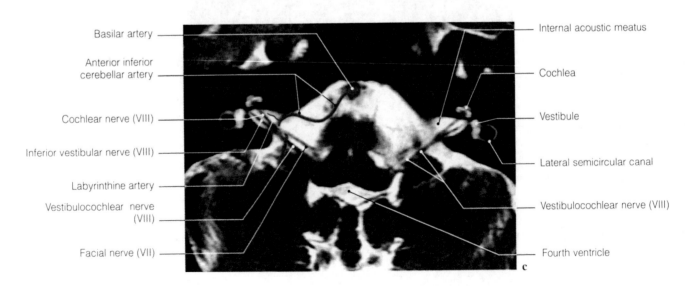

Basilar artery —
Anterior inferior cerebellar artery —
Cochlear nerve (VIII) —
Inferior vestibular nerve (VIII) —
Labyrinthine artery —
Vestibulocochlear nerve (VIII) —
Facial nerve (VII) —

— Internal acoustic meatus
— Cochlea
— Vestibule
— Lateral semicircular canal
— Vestibulocochlear nerve (VIII)
— Fourth ventricle

c

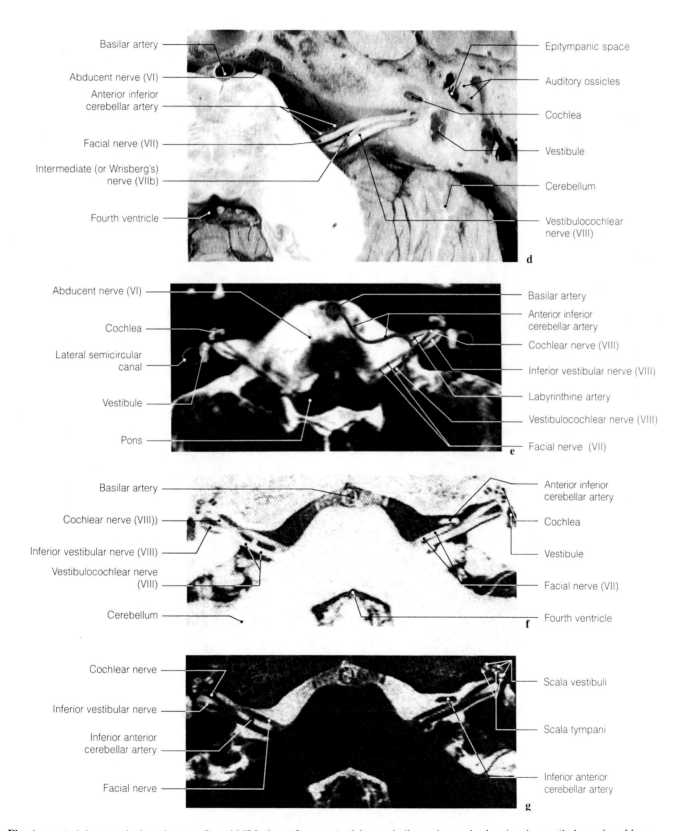

Fig. 6 a–g. Axial anatomical sections **(a, d)** and MRI views **(b, c, e–g)** of the cerebellopontine angle showing the vestibular and cochlear nerves (anatomical sections: Prof. J. P. Francke, Faculty of Medicine, Lille; MRI: **b, c, e,** Prof. Y.S. Cordoliani, Dr. J. L.Sarrazin, Hôpital du Val-de-Grâce, Paris; **f, g,** Dr. J. W. Casselman, A.Z. St Jan, Brugge)

INTERNAL EAR AND VESTIBULOCOCHLEAR PATHWAYS
ORGANS OF HEARING AND BALANCE
SPIRAL ORGAN OF CORTI, AMPULLARY CRESTS, MEMBRANOUS LABYRINTH, MODIOLUS, UTRICLE, SACCULE, VESTIBULE, SEMICIRCULAR CANALS

Anatomy, diagrams, CT and MRI views
(Pages 12-28)

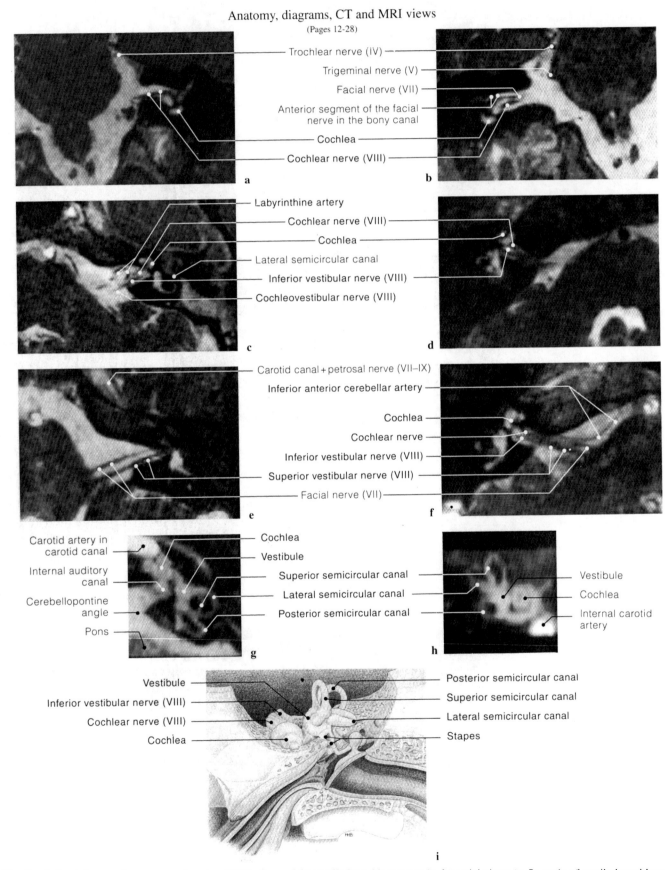

Trochlear nerve (IV)
Trigeminal nerve (V)
Facial nerve (VII)
Anterior segment of the facial nerve in the bony canal
Cochlea
Cochlear nerve (VIII)

a   b

Labyrinthine artery
Cochlear nerve (VIII)
Cochlea
Lateral semicircular canal
Inferior vestibular nerve (VIII)
Cochleovestibular nerve (VIII)

c   d

Carotid canal + petrosal nerve (VII–IX)
Inferior anterior cerebellar artery
Cochlea
Cochlear nerve
Inferior vestibular nerve (VIII)
Superior vestibular nerve (VIII)
Facial nerve (VII)

e   f

Carotid artery in carotid canal
Internal auditory canal
Cerebellopontine angle
Pons

Cochlea
Vestibule
Superior semicircular canal
Lateral semicircular canal
Posterior semicircular canal

Vestibule
Cochlea
Internal carotid artery

g   h

Vestibule
Inferior vestibular nerve (VIII)
Cochlear nerve (VIII)
Cochlea

Posterior semicircular canal
Superior semicircular canal
Lateral semicircular canal
Stapes

i

Fig. 7 a–i. Frontal magnetic resonance imaging (MRI) views of the vestibulocochlear nerve **(a, b)**; axial views **(c–f)**; study of vestibulocochlear pathways **(g–i)** of antro-attical passages and of the ossicles and vestibulocochlear pathways. (MRI: Dr. J. W. Casselman, A.Z. St Jan, Brugge)

The internal ear includes the *bony labyrinth* and the *membranous labyrinth.*

The vestibular and acoustic nerve pathways arise from the membranous labyrinth.

The bony labyrinth consists of three parts :
– a media chamber, the vestibule,
– the semicircular canals in a posterior position,
– the cochlea, located anteriorly.

The internal acoustic meatus also forms part of the bony labyrinth (Fig. 8. b).

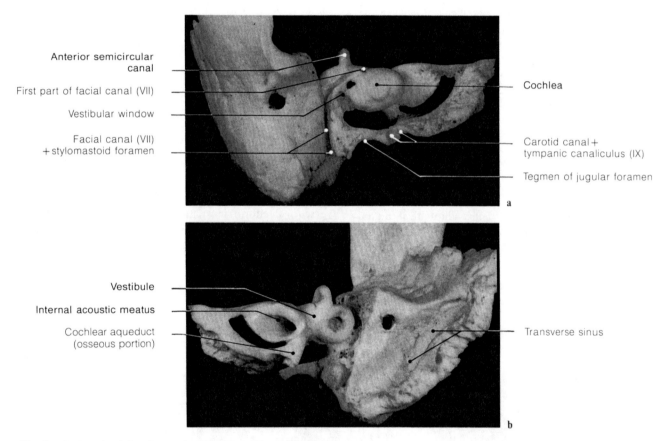

Anterior semicircular canal

First part of facial canal (VII)

Vestibular window

Facial canal (VII) + stylomastoid foramen

Cochlea

Carotid canal + tympanic canaliculus (IX)

Tegmen of jugular foramen

a

Vestibule

Internal acoustic meatus

Cochlear aqueduct (osseous portion)

Transverse sinus

b

Fig. 8 a, b. Anterior (a) and posterior (b) views of bony labyrinth

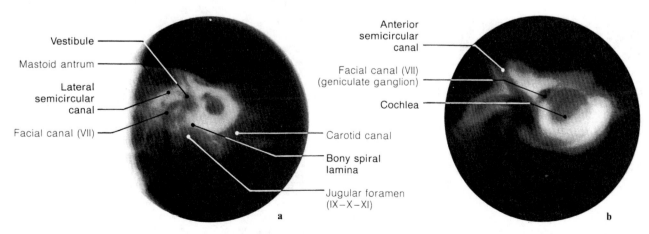

Vestibule

Mastoid antrum

Lateral semicircular canal

Facial canal (VII)

Carotid canal

Bony spiral lamina

Jugular foramen (IX–X–XI)

a

Anterior semicircular canal

Facial canal (VII) (geniculate ganglion)

Cochlea

b

Fig. 9 a, b. Tomograms of semicircular canals, vestibule and facial canal (anatomic specimen)

Anatomy

Course – Relations

The *cochlear ganglion* occupies the extent of the spiral cochlear canal.

The ramifications of origin of the vestibulocochlear nerve join the cochlear ganglion via the canaliculi of the secondary spiral lamina.

The axons of the cells of the cochlear ganglion constitute the fibres of the cochlear nerve. This nerve reaches the pons in the lateral part of the medullopontine sulcus. It ends in the ventral and dorsal cochlear nuclei.

The *ganglion of Scarpa* is situated in the fundus of the internal acoustic meatus (Fig. 10, 11). The axonal prolongations of its cells conduct sensations arising in the saccule, utricle and the ampullae of the semicircular canals.

These axons form the fibers of the vestibular nerve, which enters the pons at the same site as the cochlear nerve and medial to it. It terminates in the nuclei of the vestibular region of the floor of the fourth ventricle.

Possible causes

- Acoustic neurinoma,
- extensive cholesteatoma of the antro-adito-attical region expanding inwards and capable of damaging either the vestibule or the cochlear region,
- otosclerosis of the base of stapes, stage 5, penetrating the vestibule and the first turn of the cochlear spiral,
- fracture of the bony labyrinth, especially of the internal acoustic meatus,
- spreading fracture with disjunction of the lambdoid suture.

Examination

- Radiologic and computed tomographic (CT) studies of the internal acoustic meatuses in symmetric, intraorbital, frontal survey view,
- radiographic or CT study of the internal acoustic meatus in the long axis of the petrous bone in unilateral Chaussé IV view,
- study of the internal acoustic meatus in Pöschl-Meyer view,
- radiography in Stenvers view,
- study in 40° opposed transorbital view of François and Barrois,
- tomographic study in prestudied inclined sagittal view of Cornélis.

These views display the internal acoustic meatus, its transverse (or falciform) crest, the cochlea, vestibule, semicircular canals, chain of ossicles and the antro-adito-attical region (Fig. 27–36).

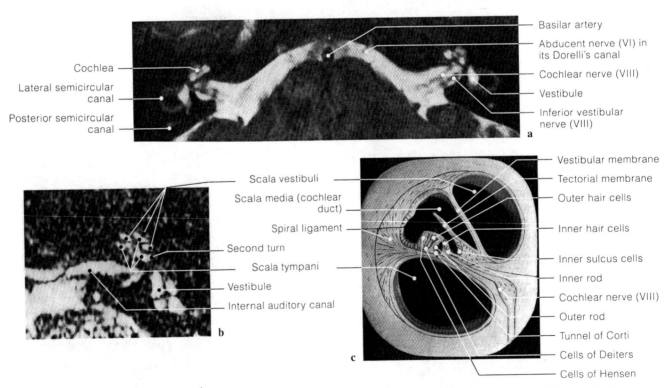

Fig. 10 a–j. Magnetic resonance (MRI) view and section through the cochlea showing the detail of spiral organ of Corti (**a–c**); diagrams, tomography, MRI, computed tomography (CT) and anatomic sections showing the vestibulocochlear nerve and the cavities of the internal ear (**d–j**). (MRI: Dr. J. W. Casselman, A.Z. St Jan, Brugge)

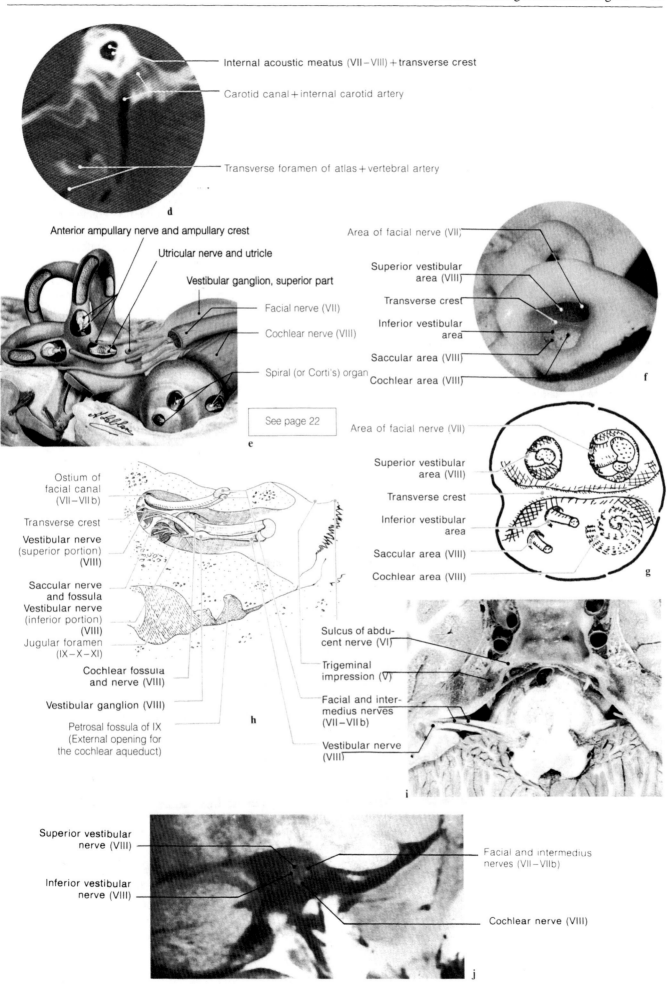

Internal acoustic meatus (VII–VIII)+transverse crest

Carotid canal+internal carotid artery

Transverse foramen of atlas+vertebral artery

d

Anterior ampullary nerve and ampullary crest

Utricular nerve and utricle

Vestibular ganglion, superior part

Facial nerve (VII)

Cochlear nerve (VIII)

Spiral (or Corti's) organ

e

See page 22

Area of facial nerve (VII)

Superior vestibular area (VIII)

Transverse crest

Inferior vestibular area

Saccular area (VIII)

Cochlear area (VIII)

f

Area of facial nerve (VII)

Superior vestibular area (VIII)

Transverse crest

Inferior vestibular area

Saccular area (VIII)

Cochlear area (VIII)

g

Ostium of facial canal (VII–VIIb)

Transverse crest

Vestibular nerve (superior portion) (VIII)

Saccular nerve and fossula

Vestibular nerve (inferior portion) (VIII)

Jugular foramen (IX–X–XI)

Cochlear fossula and nerve (VIII)

Vestibular ganglion (VIII)

Petrosal fossula of IX (External opening for the cochlear aqueduct)

h

Sulcus of abducent nerve (VI)

Trigeminal impression (V)

Facial and intermedius nerves (VII–VIIb)

Vestibular nerve (VIII)

i

Superior vestibular nerve (VIII)

Inferior vestibular nerve (VIII)

Facial and intermedius nerves (VII–VIIb)

Cochlear nerve (VIII)

j

The **membranous labyrinth** is a system of interrelated cavities with membranous walls. The posterior labyrinth is involved in balance. It comprises the saccule, the utricle, the semicircular canals, and the endolymphatic duct. The anterior labyrinth consists in the cochlear duct, and is responsible for audition.

These cavities communicate by means of canals filled endolymph. The semicircular canals and the cochlear duct follow the bony cavities (Fig. 18. a–d).

A second compartment the perilymphatic space filled with fluid is located between the osseous labyrinth and the membranous labyrinth: (Fig. 19. a–d).

The organ of Corti:

- It is a sensorineural organ containing the auditory receptors. It rests on the basilar membrane between two sulci: the external spiral sulcus and the internal spiral sulcus.
- sensory epithelium in the spiral organ of Corti is composed of three rows of outer hair cells with stereocilia (Fig. 12. e).

These cells may be contractile.

There is only one row of inner hair cells.

The **tectorial membrane of cochlear duct** is composed of a gelatinous layer and a superficial fibrous layer. It can be divided into three segments: the internal segment, the medial segment, and the external segment (Fig. 12; 13).

The **utricle** is a long, oval vesicle solidly attached with utricular nervous fibers and conjunctive tissue.
– The macula represents the sensory area of the utricle. It is located on the anterior part of the floor, facing the semioval fossula, in a horizontal plane.
– The utricular branch of the endolymphatic duct opens immediately behind the macula (Fig. 16. b, c, e).
– The apertures of the semicircular canals in the utricle are divided into two groups:
• the ampullary orifice of the posterior canal opens with the non-ampullary orifice of the lateral canal;
• the ampullary orifices of both the anterior and lateral canals open in the ceiling of the anterior extremity (Fig. 16. e).

The **saccule** is a rounded vesicle resting on the floor of the vestibule. From its inferior posterior pole stems the canalis reuniens, which connects the saccule to the cochlear duct. The posterior inferior end of the saccule bears the saccular branch of the endolymphatic duct (Fig. 16. c, e).

The macula of saccule is located on the medial face, in the vertical plane.

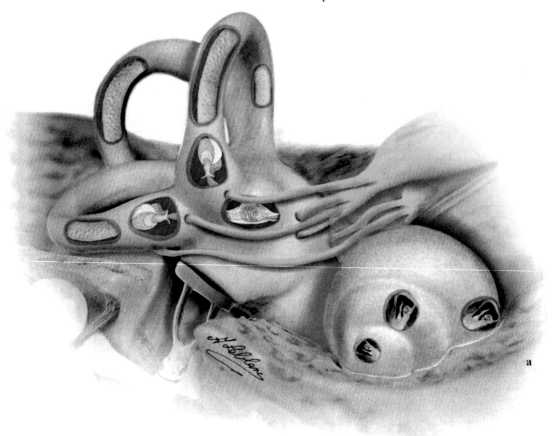

See caption on page 22

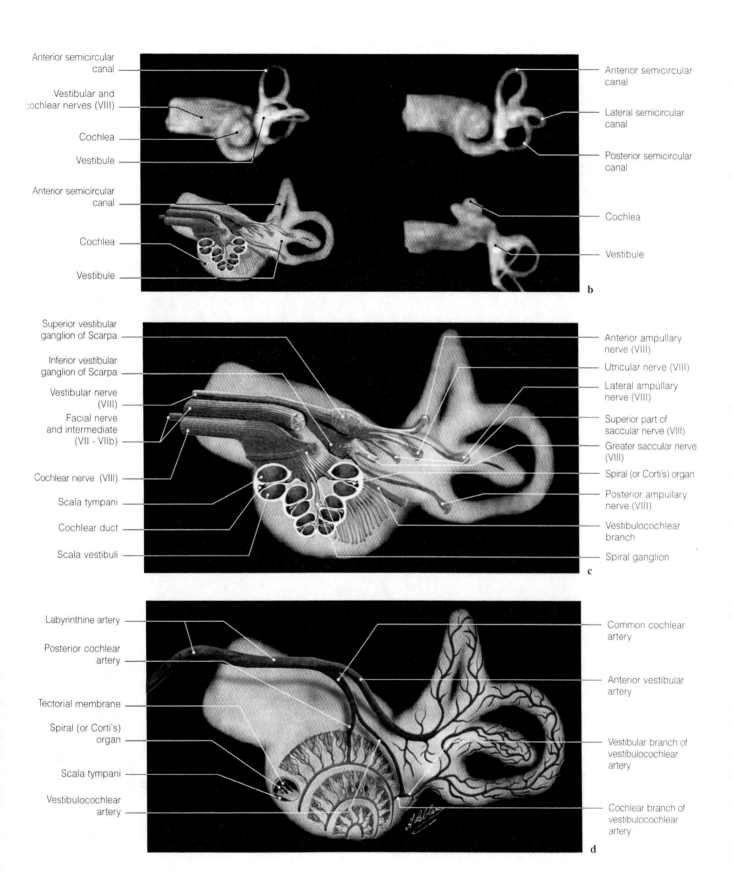

Fig. 11 a–d. Diagram of fenestrations to evidence the organs of hearing and balance **(a)**; diagram of nerves and arteries superimposed on CT views of the cochlea and the semicircular canals **(b–d)**. (CT: Prof. Y. S. Cordoliani, Dr. J. L. Sarrazin, Hôpital du Val-de-Grâce, Paris)

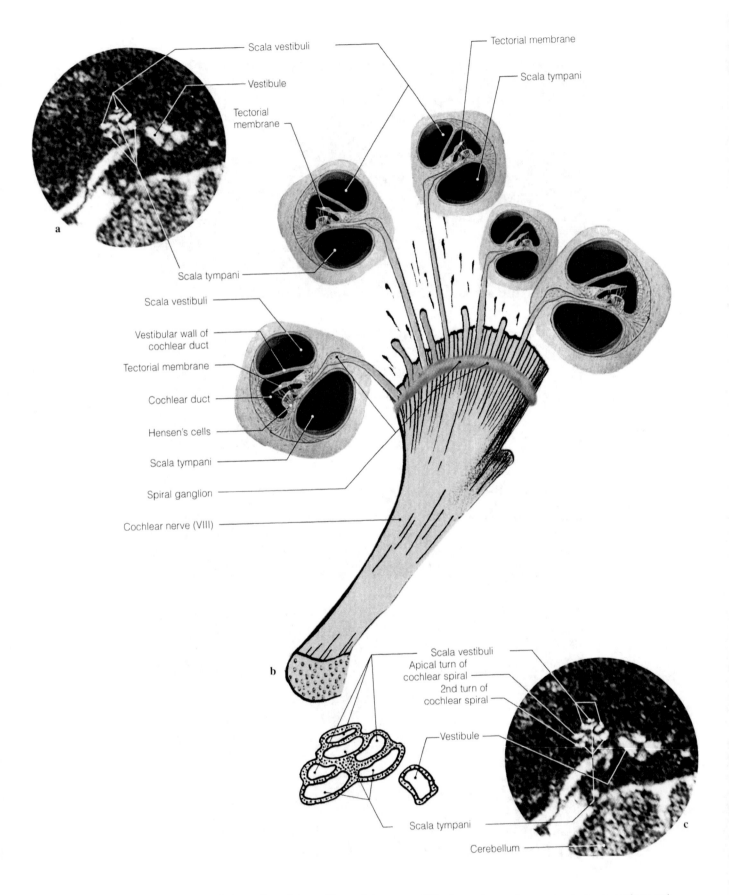

Fig. 12 a–g. Diagrams of the spiral ganglion of the cochlea and the organ of Corti superimposed on anatomical preparations and correlated to MRI views

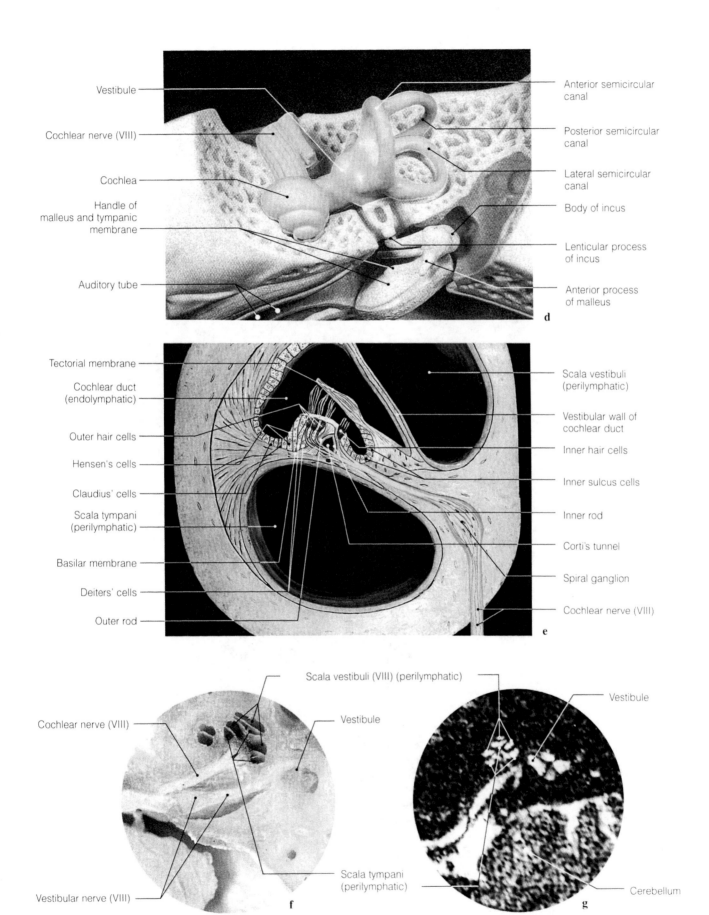

Vestibule

Cochlear nerve (VIII)

Cochlea

Handle of malleus and tympanic membrane

Auditory tube

Anterior semicircular canal

Posterior semicircular canal

Lateral semicircular canal

Body of incus

Lenticular process of incus

Anterior process of malleus

d

Tectorial membrane

Cochlear duct (endolymphatic)

Outer hair cells

Hensen's cells

Claudius' cells

Scala tympani (perilymphatic)

Basilar membrane

Deiters' cells

Outer rod

Scala vestibuli (perilymphatic)

Vestibular wall of cochlear duct

Inner hair cells

Inner sulcus cells

Inner rod

Corti's tunnel

Spiral ganglion

Cochlear nerve (VIII)

e

Scala vestibuli (VIII) (perilymphatic)

Cochlear nerve (VIII)

Vestibule

Vestibular nerve (VIII)

Scala tympani (perilymphatic)

f

Vestibule

Cerebellum

g

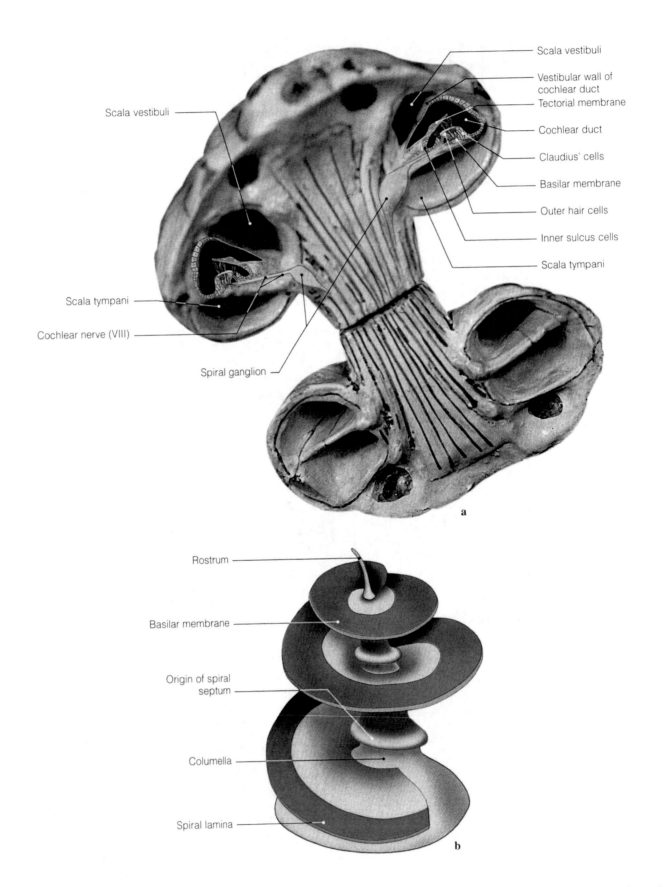

Scala vestibuli

Scala vestibuli

Vestibular wall of
cochlear duct

Tectorial membrane

Cochlear duct

Claudius' cells

Basilar membrane

Outer hair cells

Inner sulcus cells

Scala tympani

Scala tympani

Cochlear nerve (VIII)

Spiral ganglion

a

Rostrum

Basilar membrane

Origin of spiral
septum

Columella

Spiral lamina

b

Fig. 13 a–e. Diagrams of the components of the organ of Corti (**a, c, d**) and of the bony spiral lamina of the cochlea: endo- and extra-cochlear views (**b, e**)

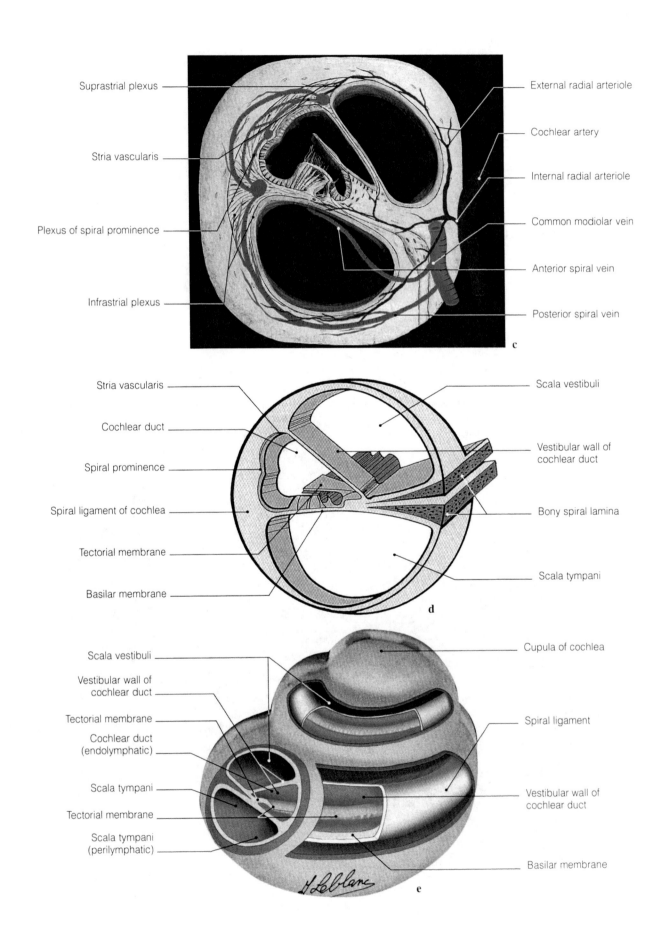

Suprastrial plexus — External radial arteriole

Stria vascularis — Cochlear artery

— Internal radial arteriole

Plexus of spiral prominence — Common modiolar vein

— Anterior spiral vein

Infrastrial plexus — Posterior spiral vein

c

Stria vascularis — Scala vestibuli

Cochlear duct

Spiral prominence — Vestibular wall of cochlear duct

Spiral ligament of cochlea — Bony spiral lamina

Tectorial membrane

Basilar membrane — Scala tympani

d

Scala vestibuli — Cupula of cochlea

Vestibular wall of cochlear duct

Tectorial membrane — Spiral ligament

Cochlear duct (endolymphatic)

Scala tympani — Vestibular wall of cochlear duct

Tectorial membrane

Scala tympani (perilymphatic) — Basilar membrane

e

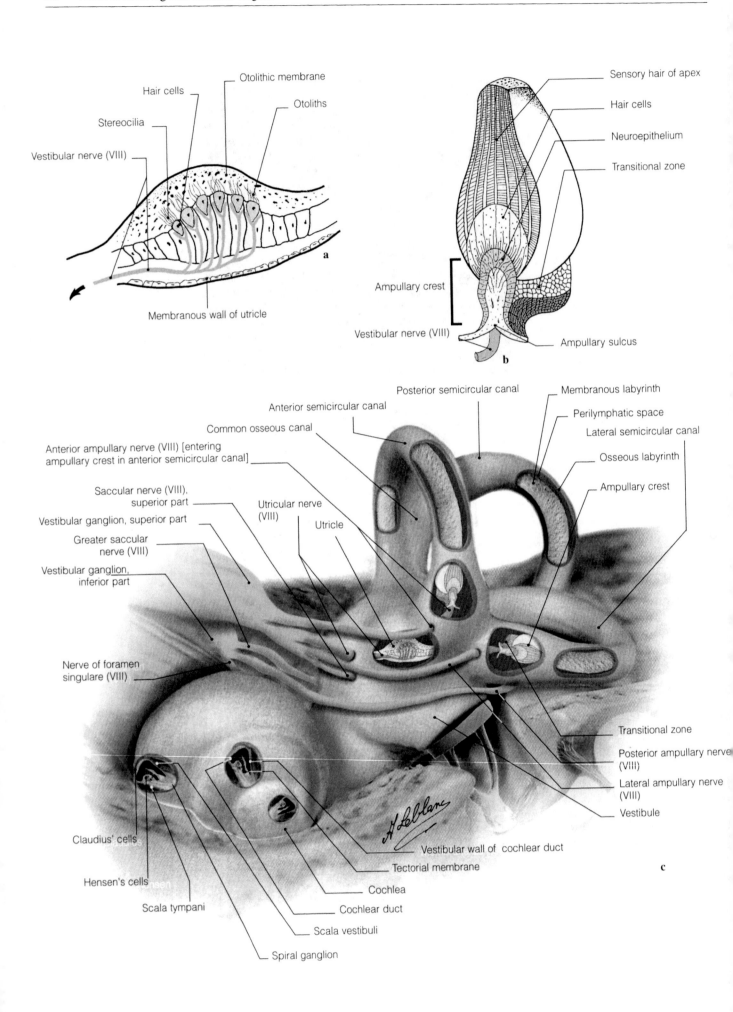

Hair cells

Otolithic membrane

Otoliths

Stereocilia

Vestibular nerve (VIII)

Membranous wall of utricle

a

Sensory hair of apex

Hair cells

Neuroepithelium

Transitional zone

Ampullary crest

Vestibular nerve (VIII)

Ampullary sulcus

b

Posterior semicircular canal

Anterior semicircular canal

Membranous labyrinth

Common osseous canal

Perilymphatic space

Lateral semicircular canal

Anterior ampullary nerve (VIII) [entering ampullary crest in anterior semicircular canal]

Osseous labyrinth

Ampullary crest

Saccular nerve (VIII), superior part

Utricular nerve (VIII)

Vestibular ganglion, superior part

Utricle

Greater saccular nerve (VIII)

Vestibular ganglion, inferior part

Nerve of foramen singulare (VIII)

Transitional zone

Posterior ampullary nerve (VIII)

Lateral ampullary nerve (VIII)

Vestibule

Claudius' cells

Vestibular wall of cochlear duct

Hensen's cells

Tectorial membrane

Scala tympani

Cochlea

Cochlear duct

Scala vestibuli

Spiral ganglion

c

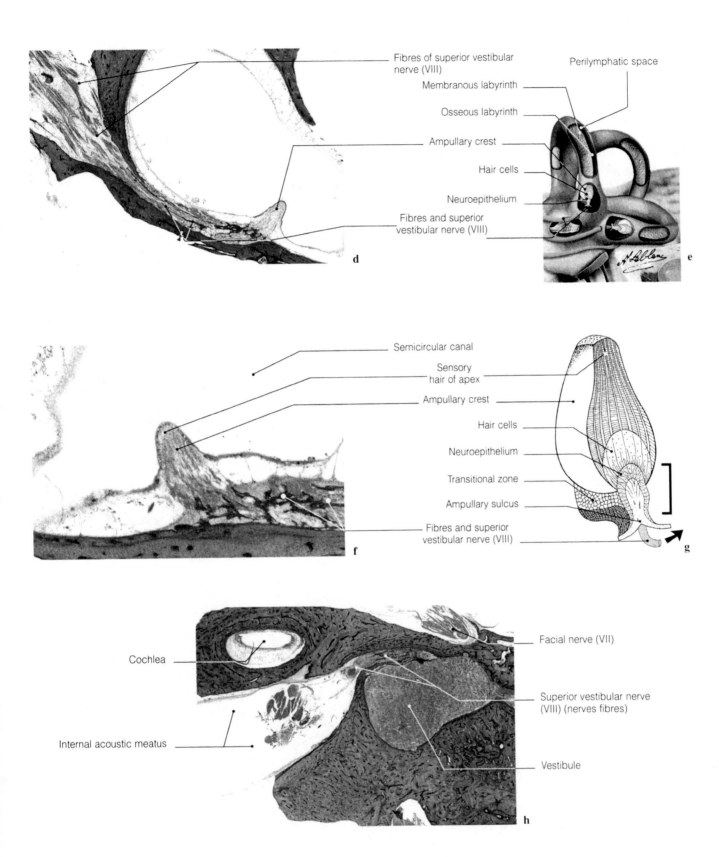

Fibres of superior vestibular nerve (VIII)

Membranous labyrinth

Osseous labyrinth

Ampullary crest

Hair cells

Neuroepithelium

Fibres and superior vestibular nerve (VIII)

Perilymphatic space

d

e

Semicircular canal

Sensory hair of apex

Ampullary crest

Hair cells

Neuroepithelium

Transitional zone

Ampullary sulcus

Fibres and superior vestibular nerve (VIII)

f

g

Cochlea

Facial nerve (VII)

Superior vestibular nerve (VIII) (nerves fibres)

Internal acoustic meatus

Vestibule

h

Fig. 14 a–h. Diagrams and imaging (cochlear implant) of the utricle, the ampullary crests and the vestibular nerve

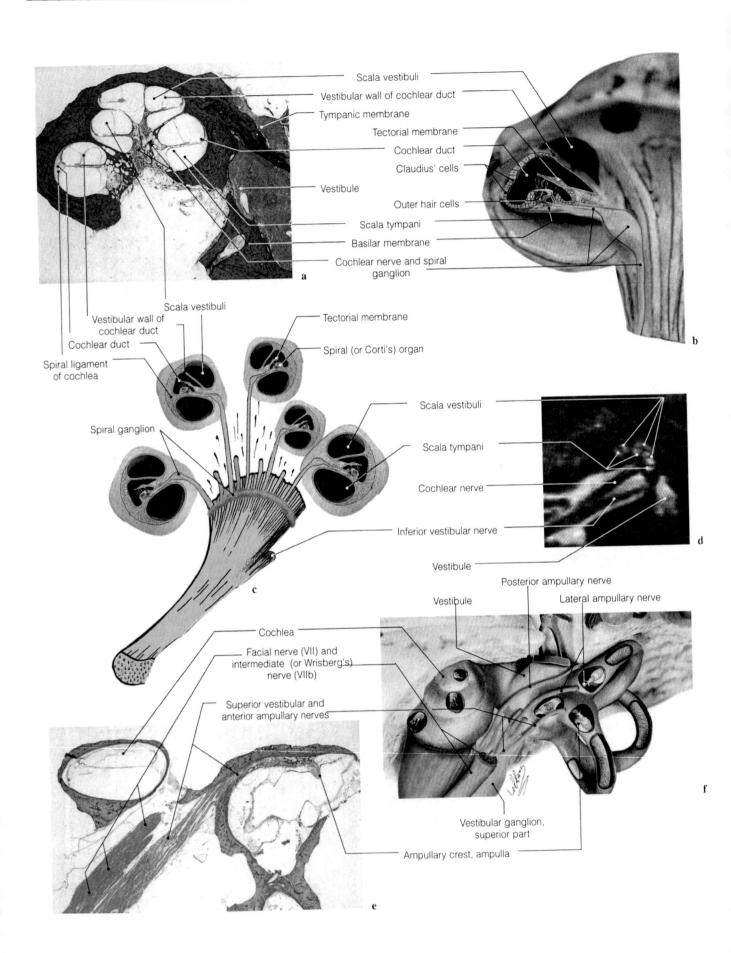

Scala vestibuli

Vestibular wall of cochlear duct

Tympanic membrane

Tectorial membrane

Cochlear duct

Claudius' cells

Vestibule

Outer hair cells

Scala tympani

Basilar membrane

Cochlear nerve and spiral ganglion

a

b

Scala vestibuli

Vestibular wall of cochlear duct

Cochlear duct

Spiral ligament of cochlea

Tectorial membrane

Spiral (or Corti's) organ

Spiral ganglion

Scala vestibuli

Scala tympani

Cochlear nerve

Inferior vestibular nerve

Vestibule

c

d

Posterior ampullary nerve

Vestibule

Lateral ampullary nerve

Cochlea

Facial nerve (VII) and intermediate (or Wrisberg's) nerve (VIIb)

Superior vestibular and anterior ampullary nerves

Vestibular ganglion, superior part

Ampullary crest, ampulla

e

f

The modiolus

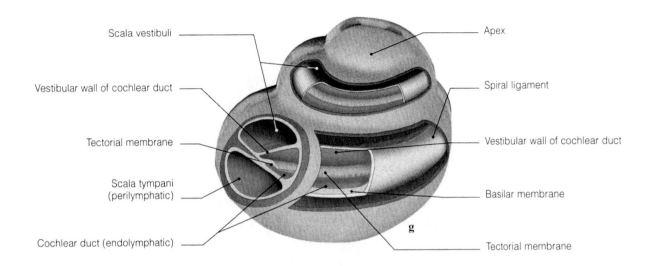

Scala vestibuli

Apex

Vestibular wall of cochlear duct

Spiral ligament

Tectorial membrane

Vestibular wall of cochlear duct

Scala tympani
(perilymphatic)

Basilar membrane

g

Cochlear duct (endolymphatic)

Tectorial membrane

For extracochlear course of bony spiral lamina see **Fig. 13. b**

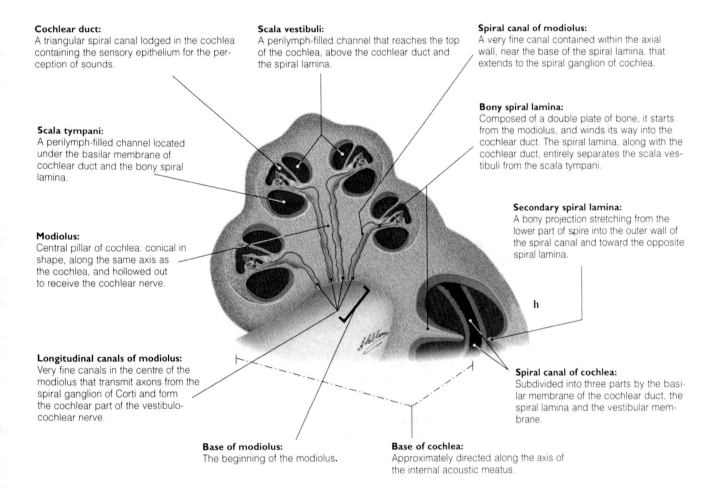

Cochlear duct:
A triangular spiral canal lodged in the cochlea containing the sensory epithelium for the perception of sounds.

Scala vestibuli:
A perilymph-filled channel that reaches the top of the cochlea, above the cochlear duct and the spiral lamina.

Spiral canal of modiolus:
A very fine canal contained within the axial wall, near the base of the spiral lamina, that extends to the spiral ganglion of cochlea.

Scala tympani:
A perilymph-filled channel located under the basilar membrane of cochlear duct and the bony spiral lamina.

Bony spiral lamina:
Composed of a double plate of bone, it starts from the modiolus, and winds its way into the cochlear duct. The spiral lamina, along with the cochlear duct, entirely separates the scala vestibuli from the scala tympani.

Modiolus:
Central pillar of cochlea: conical in shape, along the same axis as the cochlea, and hollowed out to receive the cochlear nerve.

Secondary spiral lamina:
A bony projection stretching from the lower part of spire into the outer wall of the spiral canal and toward the opposite spiral lamina.

h

Longitudinal canals of modiolus:
Very fine canals in the centre of the modiolus that transmit axons from the spiral ganglion of Corti and form the cochlear part of the vestibulocochlear nerve.

Spiral canal of cochlea:
Subdivided into three parts by the basilar membrane of the cochlear duct, the spiral lamina and the vestibular membrane.

Base of modiolus:
The beginning of the modiolus.

Base of cochlea:
Approximately directed along the axis of the internal acoustic meatus.

Fig. 15. a-h. Diagram showing the relationship between modiulus and cochlear nerve; bony spiral lamina and the connections between the cochlear canal and the scalae tympani and vestibuli (**a-d ; g,h**); MRI views correlated to a diagram for the ampullar crests, with their corresponding utricular nerve (**e, f**)

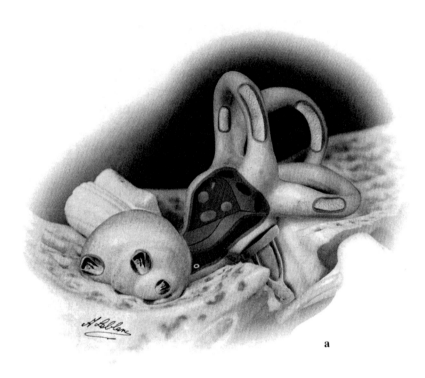

a

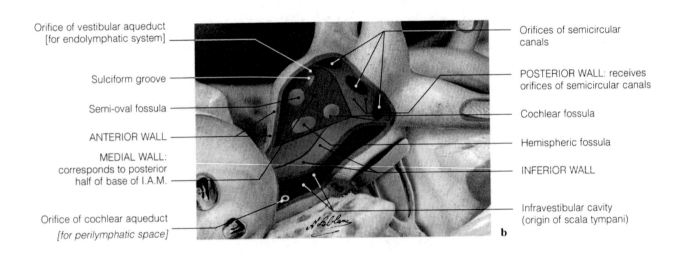

Orifice of vestibular aqueduct [for endolymphatic system]

Sulciform groove

Semi-oval fossula

ANTERIOR WALL

MEDIAL WALL: corresponds to posterior half of base of I.A.M.

Orifice of cochlear aqueduct [for perilymphatic space]

Orifices of semicircular canals

POSTERIOR WALL: receives orifices of semicircular canals

Cochlear fossula

Hemispheric fossula

INFERIOR WALL

Infravestibular cavity (origin of scala tympani)

b

Fig. 16 a–e. Fenestration of the vestibule to show the openings, of the semicircular canals, the cochlear and vestibular aqueducts in the perforated bony walls (a, b); the same fenestrated vestibule showing the position of the utricle, the saccule and the endolymphatic duct (c–e)

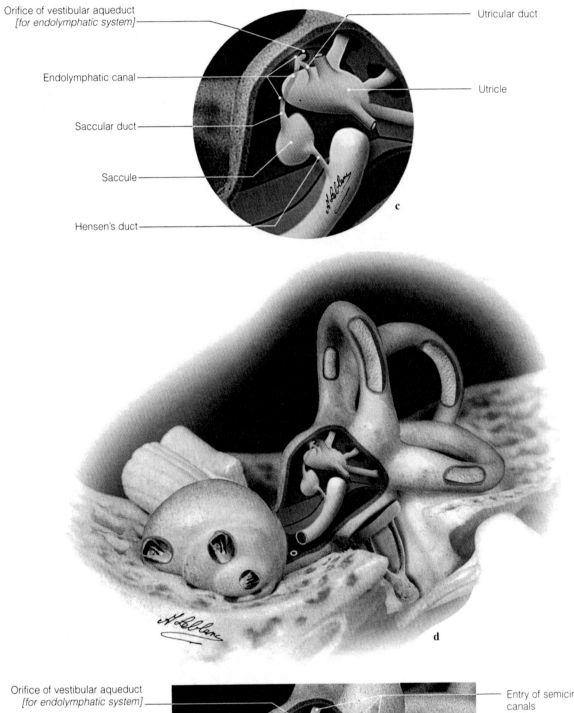

Orifice of vestibular aqueduct
[for endolymphatic system]

Endolymphatic canal

Saccular duct

Saccule

Hensen's duct

Utricular duct

Utricle

c

d

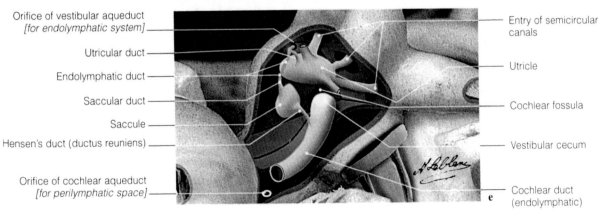

Orifice of vestibular aqueduct
[for endolymphatic system]

Utricular duct

Endolymphatic duct

Saccular duct

Saccule

Hensen's duct (ductus reuniens)

Orifice of cochlear aqueduct
[for perilymphatic space]

Entry of semicircular
canals

Utricle

Cochlear fossula

Vestibular cecum

Cochlear duct
(endolymphatic)

e

Semicircular canals

They are composed of three membranous ducts that open into the utricle via five orifices (Fig. 16. a–e).

Each duct is contained in a semicircular canal, but only occupies a little over a fourth of its diameter, and is affixed to the bony canal by its outer wall.

The ducts each possess their own membrane. A basal membrane rests upon it and carries the epithelial cells (Fig. 16. a–e).

The dilated ends of the semicircular ducts form ampullae, all three of which are very near the utricle.

– Each ampulla features a sulcus. This fold, located on the medial part of the ampulla, corresponds to a localized thickening of the membrane called the ampullary crest. Nervous fibres enter the ampullae through this sulcus.

- The ampullary crests are covered with a neuroepithelium of cells with one kinocilia and smaller stereocilia (Fig. 14. b, c; 15. e, f).

The stereocilia and kinocilia are both located on the same side in an ampulla.

In the lateral semicircular duct, the kinocilia are located on the wall near the utricle, while the opposite is true of the posterior and anterior ampullae.

The cupula is a mass that rests on the neuroepithelium affixed to the walls of the ampulla. It seals the duct containing the endolabyrinthic fluids. Acceleration causes both the cupula and the cilia to undergo angular movements. The sterocilia and kinocilia, immersed in the fluids, analyse their deformation.

ENDOLYMPHATIC SYSTEM
PERILYMPHATIC SPACE

THE ENDOLYMPHATIC SYSTEM

(Pages 29, 30)

Endolymphatic sac and duct

The **endolymphatic duct** is formed by the reunion of two canaliculi that come from the saccule and the utricle.

The utricular segment merges with the utricle through a very small fissure.

The endolymphatic duct first presents an enlarged intravestibular portion called the sinus. Its diameter decreases when it reaches the isthmus and penetrates the aqueduct of the vestibule, and finally becomes larger again. The duct is covered with connective tissue along most of its length. The type of epithelium varies with the portion of the duct.

The **endolymphatic sac** continues and ends the endolymphatic duct. It constitutes a true intracranial extension of the membranous labyrinth of approximately 10 mm width,

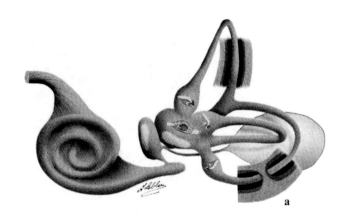

overlapping the ungual fossa to line the dura mater (Fig. 18 a–d)

THE PERILYMPHATIC SPACES

(Pages 29-31)

Cochlear aqueduct and perilymphatic duct

The cochlear aqueduct communicates with the perilymphatic space (Fig. 19 a–e).

Its ostium is located on the medial cochlear wall, very close to the round window and the inferior side of the spiral lamina.

It is directed toward the back, downward and to the inside. It slips under the ampulla of the posterior semicircular canal, at the level of the inferior edge of the acoustic meatus.

It ends at the lower part of the pyramid, and opens in the floor of the petrosal fossa [cavitas of Andersch's ganglion (IX)] behind the jugular foramen and before the carotid canal, next to the tympanic canaliculus (Fig. 19. b, d).

The perilymphatic duct connects the perilymphatic space with the subarachnoid space located between the arachnoid and the pia mater.

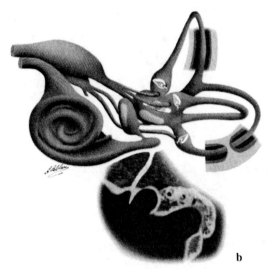

Fig. 17 a, b. Diagrams of the endolymphatic system (**a**) and the perilymphatic spaces (**b**).

ENDOLYMPHATIC SYSTEM
ENDOLYMPHATIC SAC, ENDOLYMPHATIC DUCT
VESTIBULAR AQUEDUCT
Anatomical diagrams and CT views

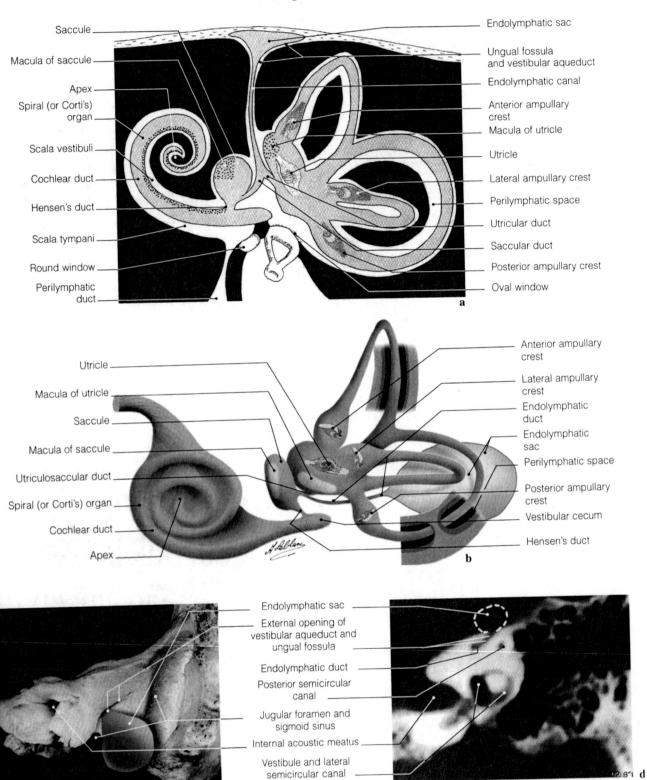

Fig. 18 a–d. Anatomical diagrams showing the endolymphatic system (**a, b**); computerized axial tomography of the vestibular aqueduct and the endolymphatic duct (**d**); photograph of the posterior area of the petrosal part of the temporal bone (**c**) showing the external opening of the vestibular aqueduct (ungual fossa) with a diagram of the endolymphatic sac (CT: Dr. J .W. Casselman, A.Z. St Jan, Brugge)

PERILYMPHATIC SPACE
PERILYMPHATIC DUCT, COCHLEAR AQUEDUCT
Anatomical diagrams and CT views

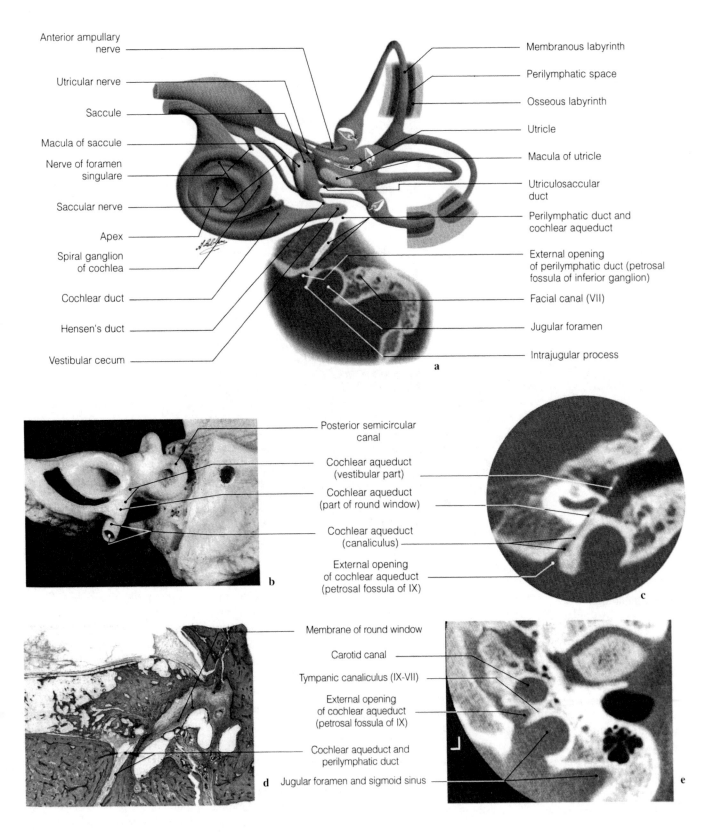

Fig. 19 a–e. Diagram showing the perilymphatic spaces (**a**); posterior anatomical view of the vestibulocochlear area (**b**); imaging of the cochlear aqueduct (**d**); CT views of the cochlear aqueduct and the external ostium of its canaliculus (**c, e**) (CT: Dr. J. W. Casselman, A.Z. St Jan, Brugge)

LABYRINTHINE VASCULATURE
ANTERIOR INFERIOR CEREBELLAR ARTERY, LABYRINTHINE ARTERY, VESTIBULAR AND COCHLEAR ARTERIES

Anatomy, diagrams, CT and MRI views

(Pages 32-40)

Vasculature of the internal ear

The vasculature of the bony inner ear is independent from that of the membranous inner ear.

Vasculature of the bony labyrinth is supplied by:

- the stylomastoid artery, a branch of the posterior auricular artery;
- the tympanic artery, inferior, a branch of the ascending pharyngeal artery;
- the subarcuate artery, that originates either directly from the anterior inferior cerebellar artery, or more particularly from the internal auditory artery.

The subarcuate artery passes through the petrosomastoid canal.

The membranous labyrinth has its own vasculature:

The labyrinthine artery or internal auditory artery: supplied by the inferior anterior cerebellar artery (i.e., the middle cerebellar artery) or directly by the basilar artery, the labyrinthine artery arrives at the back of the acoustic meatus and divides into three branches (Fig. 22; 23. b):

Fig. 20 For annotations see page 39.

* the cochlear artery,
* the anterior vestibular artery,
* the cochleovestibular artery.

The cochlear artery enters the modiolus, where it forms a spiral and supplies two types of radial arteries:

- the external radial arteriolae that run under the bony scala vestibuli and supply four main capillary networks:
- the helicotrema
- the suprastrial capillaries, which are thought to secrete most of the perilymph,
- the stria vascularis of the cochlear duct,
- the infrastrial network, which is anastomosed to the venous capillaries.

– the internal radial arteriolae supply the spiral ganglion of the cochlea and the bony spiral lamina.

The anterior vestibular artery supplies branches for the posterior sides of the utricle and saccule, and extends to the lateral and anterior semicircular canals (Fig. 22; 24. d–f).

The vestibulocochlear artery divides into two branches:

The cochlear branch irrigates the inferior fourth of the cochlear duct, and then merges with the ramus cochlearis.

The posterior vestibular branch vascularizes the macula, the saccule, the walls and the ampulla of the posterior semicircular canal, and the inferior poles of the saccule and utricle.

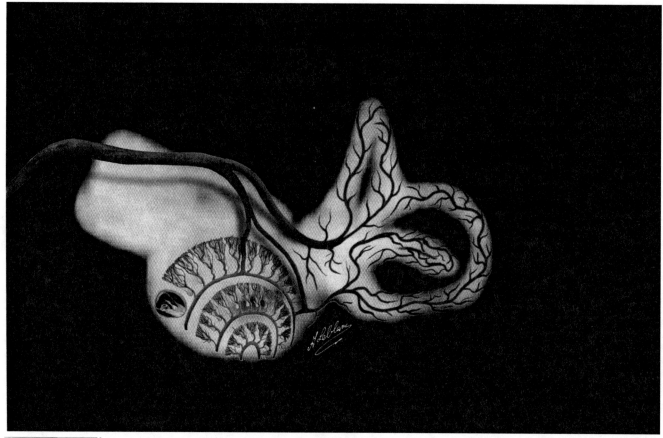

b

Annotations: see page 17

Fig. 21 a, b. Diagrams of labyrinthine vasculature (**a**), superimposed on a computed tomography (CT) view of the semicircular canals and the cochlea

Veins

There are two main venous networks:
* the plexus for the cochlear aqueduct,
* the plexus for the vestibular aqueduct.

* The plexus for the cochlear aqueduct includes:

– Veinules from the sensory areas of the vestibule:
• the posterior vestibular vein [ampulla of the posterior semicircular canal, saccule];
• the anterior vestibular vein [utricle].

– The common modiolar vein, which results from the merging of the anterior and posterior spiral veins.

• the posterior spiral vein collects blood from the spiral ganglion of cochlea and then merges with the infrastrial capillary network;
• the anterior spiral vein collects blood from the limbus spiralis, and then anastomoses with the internal radial arteriolae at that same level;

• within the modiolus, the anterior and posterior spiral veins communicate in several points.

– The vein of round window.

This plexus follows the course of the cochlear aqueduct in a separate canal, and drains into the vein of the cochlear aqueduct.

* The plexus for the vestibular aqueduct is an anastomosis of veins coming from the non-sensory areas of the vestibular labyrinth, particularly from the semicircular canals. It drains into the vein of the vestibular aqueduct, which runs in a canal parallel to the aqueduct, and finally receives veins from the endolymphatic sac.

* Both of these plexuses drain into the inferior petrosal sinus, and finally into the jugular bulb.

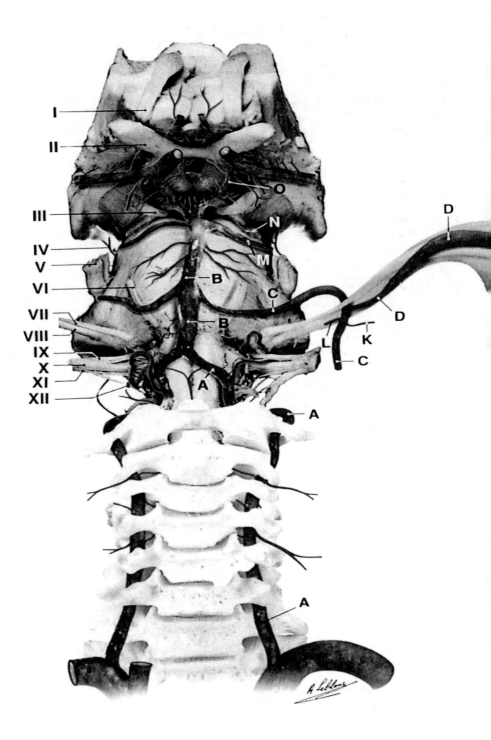

Fig. 22

VASCULARIZATION

Labyrinthine artery

A   Vertebral artery
B   Basilar artery
C   Inferior anterior cerebellar artery
D   Labyrinthine artery
E   Anterior vestibular artery
F   Common cochlear artery
G   Posterior cochlear artery

H   Cochlear branch of vestibulocochlear artery
I   Vestibulocochlear artery
J   Vestibular branch of vestibulocochlear artery
K   Subarcuate artery *[anastomoses in middle ear with branches of the external carotid artery]*
L   Recurrent artery
M   Superior cerebellar artery
N   Posterior cerebral artery
O   Posterior communicating artery

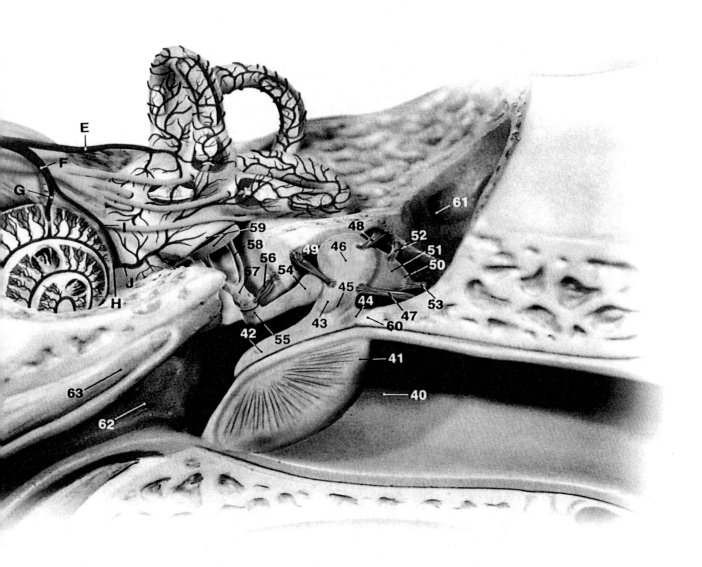

MIDDLE AND EXTERNAL EAR

40	External acoustic meatus		52	Superior ligament of incus
41	Tympanic membrane		53	Posterior ligament of incus
42	Handle of malleus		54	Long crus of incus
43	Anterior process of malleus		55	Lenticular process of incus
44	Lateral process of malleus		56	Stapedius muscle
45	Neck of malleus		57	Head of stapes
46	Head of malleus		58	Posterior crus of stapes
47	Lateral ligament of malleus		59	Oval window and base of stapes
48	Superior ligament of malleus		60	Epitympanic recess
49	Malleus muscle		61	Mastoid antrum
50	Short crus of incus		62	Auditory tube
51	Body of incus		63	Tensor tympani muscle

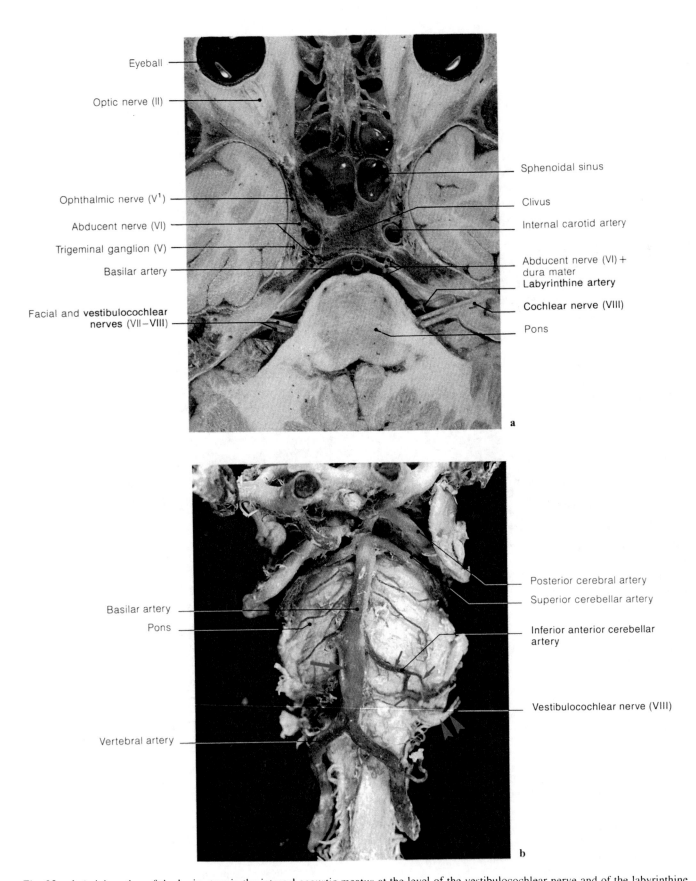

Eyeball

Optic nerve (II)

Ophthalmic nerve (V¹)

Abducent nerve (VI)

Trigeminal ganglion (V)

Basilar artery

Facial and **vestibulocochlear nerves** (VII–VIII)

Sphenoidal sinus

Clivus

Internal carotid artery

Abducent nerve (VI) + dura mater
Labyrinthine artery

Cochlear nerve (VIII)

Pons

a

Basilar artery

Pons

Vertebral artery

Posterior cerebral artery

Superior cerebellar artery

Inferior anterior cerebellar artery

Vestibulocochlear nerve (VIII)

b

Fig. 23 a–i. Axial section of the brain stem in the internal acoustic meatus at the level of the vestibulocochlear nerve and of the labyrinthine artery (**a**); dissection of the brain stem showing the anterior inferior cerebellar artery (**b**); axial anatomical section (**c**), MRI views (**d, g, h, i**) and diagrams (**e, f**) depicting the vascular relationships between vestibular, cochlear, facial and intermediate (VIII-VII-VIIb) nerves. (Dissection and anatomical sections: Prof. J. P. Francke, Faculty of Medicine, Lille; diagrams **e, f**: Prof. Y. Guerrier, *Anatomie Chirurgicale de l'os temporal de l'oreille et de la base du crâne*, Tome 1, La Simarre, 1988)

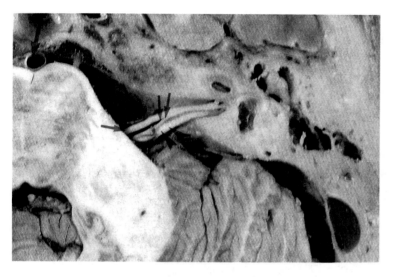

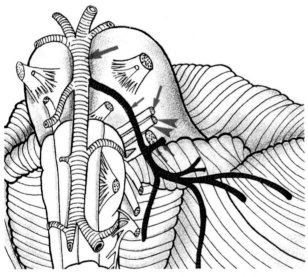

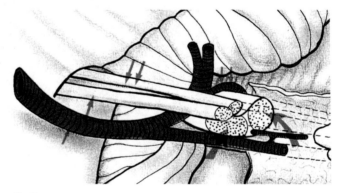

Basilar artery (*large. straight arrow*)

labyrinthine artery
(*small, double arrows*)

inferior anterior
cerebellar artery (*small arrow*)

facial nerve (VII)
(*medium, straight arrow*)

nervus intermedius (VIIb)
(*double. medium arrows*)

vestibulocochlear nerve (VIII)
(*double arrowheads*)

inferior vestibular (VIII) (*curved arrow*)

cochlear nerve (VIII) (*large arrow*)

Cochlea (*C*); vestibule (*V*)

Cochlea and cochlear nerve (VIII)

Vestibule and inferior vestibular nerve

Lateral semicircular canal

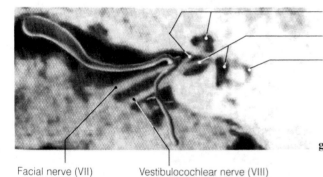

Facial nerve (VII) Vestibulocochlear nerve (VIII)

g

Labyrinthine artery

Facial nerve (VII)

Cochlear nerve
(VIII)

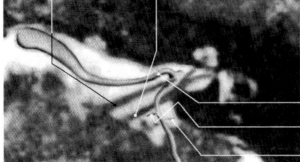

Anterior inferior cerebellar artery

Recurrent artery

Subarcuate artery [*anastomoses
in middle ear with branches
of the external carotid artery*]

Inferior vestibular
nerve (VIII)

i

(MRI: Dr. J. W. Casselman, A. Z. St Jan, Brugge)

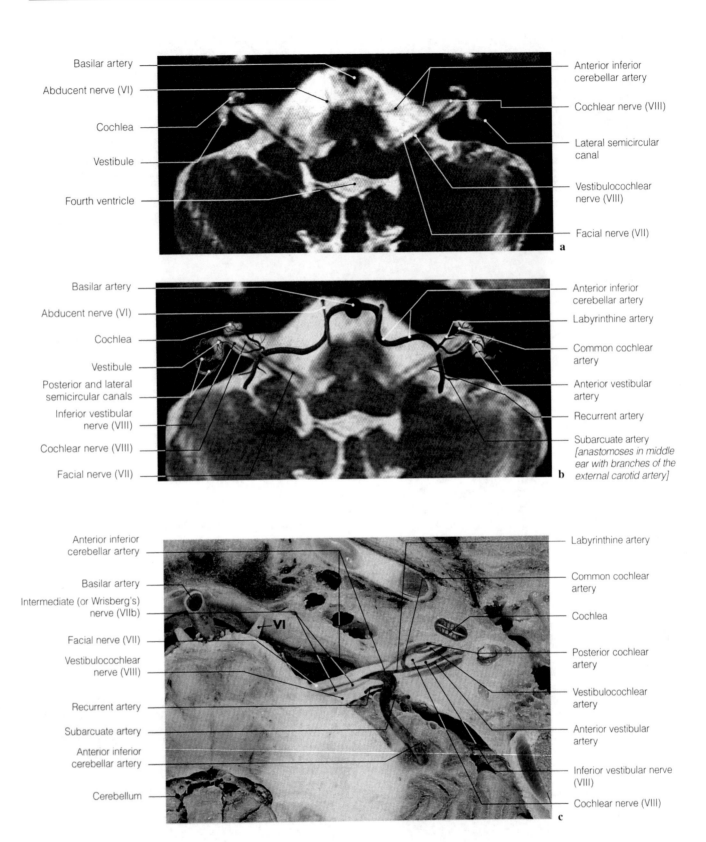

Fig. 24 a–f. MRI views of the cerebellopontine angle at the level of the acoustic and facial nerves; superimposed is a diagram of the anterior inferior cerebellar artery (**a, b**); anatomical section of the same level (**c**); diagram of arterial and venous vasculature of the internal ear (**d–f**) (MRI: Prof. Y. S. Cordoliani, Dr. J. L. Sarrazin, Hôpital du Val-de-Grâce, Paris); (anatomical section: Prof. J. P. Francke, Faculty of Medicine, Lille); (diagram (**e**): Prof. Y. Guerrier, *Anatomie Chirurgicale de l'os temporal de l'oreille et de la base du crâne*, Tome 1, La Simarre, 1988)

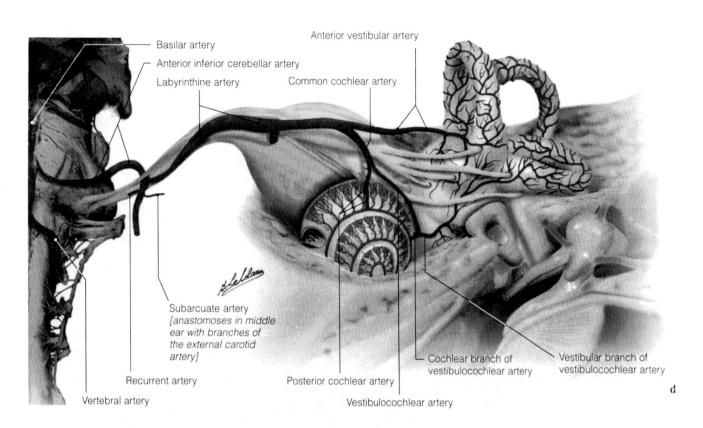

Basilar artery
Anterior inferior cerebellar artery
Labyrinthine artery
Anterior vestibular artery
Common cochlear artery

Subarcuate artery
[anastomoses in middle ear with branches of the external carotid artery]

Recurrent artery

Vertebral artery

Posterior cochlear artery

Vestibulocochlear artery

Cochlear branch of vestibulocochlear artery

Vestibular branch of vestibulocochlear artery

d

Labyrinthine artery

Anterior vestibular artery

Common cochlear artery

Posterior cochlear artery

Vestibulocochlear artery

U : Utricle

S : Saccule

e

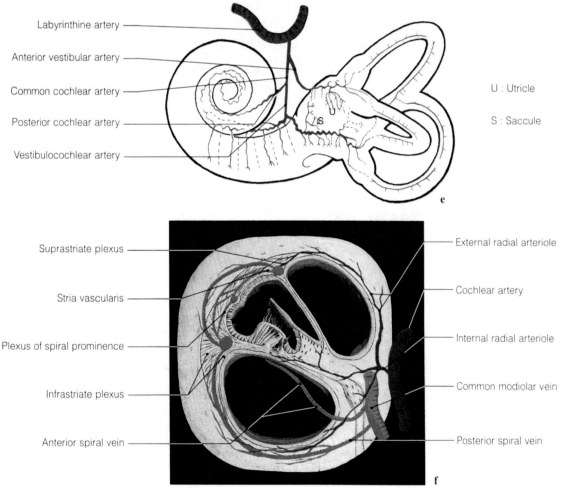

Suprastriate plexus

Stria vascularis

Plexus of spiral prominence

Infrastriate plexus

Anterior spiral vein

External radial arteriole

Cochlear artery

Internal radial arteriole

Common modiolar vein

Posterior spiral vein

f

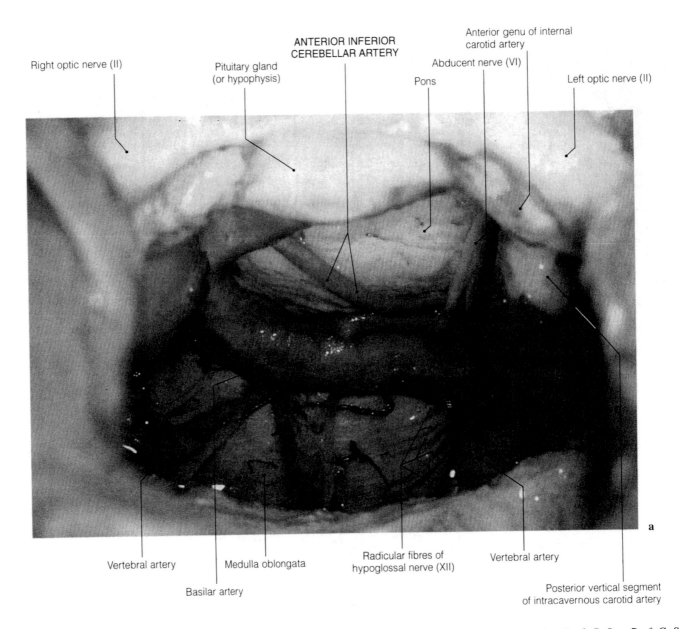

Right optic nerve (II)

Pituitary gland
(or hypophysis)

ANTERIOR INFERIOR
CEREBELLAR ARTERY

Pons

Anterior genu of internal
carotid artery

Abducent nerve (VI)

Left optic nerve (II)

Vertebral artery

Basilar artery

Medulla oblongata

Radicular fibres of
hypoglossal nerve (XII)

Vertebral artery

Posterior vertical segment
of intracavernous carotid artery

a

Fig. 25 Frontal anatomical view of the brain stem showing the basilar and anterior inferior cerebellar arteries (Prof. C. Sen, Prof. C. S. Chen, Prof. K. D. Post, *Microsurgical Anatomy of the Skull Base*, Thieme, 1997)

MIDDLE EAR, OSSICULAR CHAIN
MUSCLES, LIGAMENTS, NASOTUBAL CAVITIES, EPITYMPANIC RECESS

Anatomy, diagrams, and CT views

(Pages 41-47)

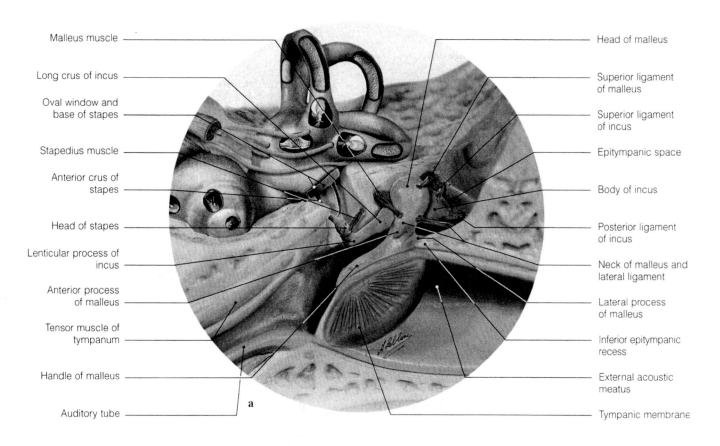

Malleus muscle

Long crus of incus

Oval window and base of stapes

Stapedius muscle

Anterior crus of stapes

Head of stapes

Lenticular process of incus

Anterior process of malleus

Tensor muscle of tympanum

Handle of malleus

Auditory tube

Head of malleus

Superior ligament of malleus

Superior ligament of incus

Epitympanic space

Body of incus

Posterior ligament of incus

Neck of malleus and lateral ligament

Lateral process of malleus

Inferior epitympanic recess

External acoustic meatus

Tympanic membrane

a

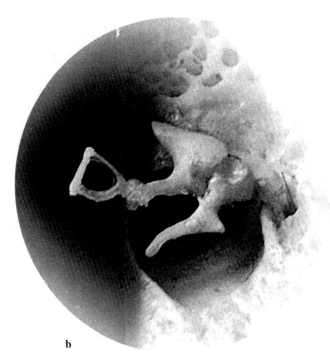

b

Fig. 26 a, b. Diagram of the auditory ossicles and their ligaments (**a**); anatomical view of the ossicular chain (**b**)

Middle ear (organ of transmission)

Anatomy

The **middle ear** is a long cavity containing air and formed of three parts:
- the tympanic compartment,
- the auditory tube,
- the mastoid cavities.

The tympanic compartment, hollowed out in the temporal bone, is separated from the external ear by the tympanic membrane and from the internal ear by
- the vestibular window above, corresponding to the vestibule,
- the cochlear window below, corresponding to the scala tympani of the cochlea.

The middle ear communicates with the rhinopharynx via the auditory tube. It is occupied by the chain of auditory ossicles
- the malleus, incus and stapes – which connect the tympanum with the vestibular window (Fig. 29–38).

The **malleus** has a head, a neck, a handle and two processes, one anterior and one lateral (Fig. 31.a–g). The head of the malleus is joined to the body of the incus at the incudomallear articulation.

The **incus** is situated behind the malleus and has a body and two limbs, long and short. It is situated in the attic and its body is flattened lateromedially, its articular surface is adapted to the articular surface of the head of the malleus.

The *short (upper or horizontal) limb* is squat, thick and of a flattened cone shape; its posterior end rests against the notch situated at the antero-inferior angle of the ostium of the *aditus ad antrum.*

The *long (lower) limb* is more slender and longer than the former and initially descends almost vertically behind and medial to the handle of the malleus. Its lower end bends inwards to terminate in a rounded tubercle: the *lenticular process* which articulates with the stapes.

The **stapes** is situated medial to the incus and extends almost horizontally from the lenticular process to the vestibular (oval) window.

It has a head, a platelike base and two limbs. Laterally, the head is hollowed by a glenoid cavity which articulates with the lenticular process of the incus. The stapes is an oval membrane connected with the vestibular window.

The limbs of the stapes are two: anterior and posterior.

Connections of the ossicles: The ossicles are interconnected by two articulations: the incudomallear and incudostapedial joints.

The **motor muscles of the ossicles** are two in number: the stapedius and the tensor tympani.

Fig. 27 a, b. Computed tomography (CT) of the antro-adito-attical passage and of the incus and base of the stapes; diagram of the middle ear withe the incudomallear and incudostapedial articulations

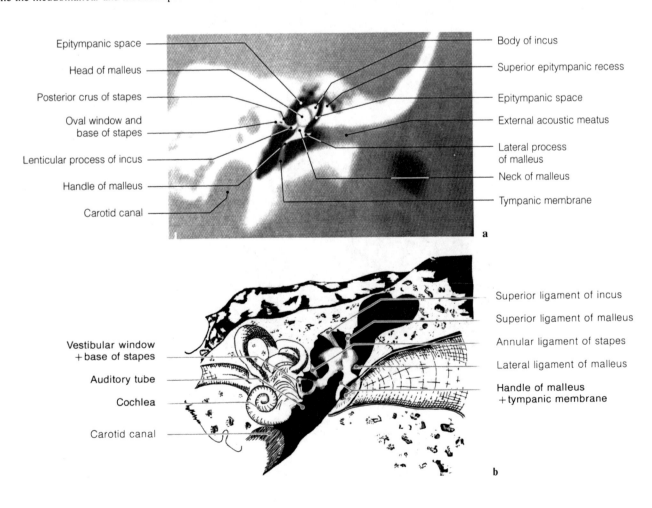

Epitympanic space — Body of incus

Head of malleus — Superior epitympanic recess

Posterior crus of stapes — Epitympanic space

Oval window and base of stapes — External acoustic meatus

Lenticular process of incus — Lateral process of malleus

Handle of malleus — Neck of malleus

Carotid canal — Tympanic membrane

a

Vestibular window + base of stapes — Superior ligament of incus

Auditory tube — Superior ligament of malleus

Cochlea — Annular ligament of stapes

Carotid canal — Lateral ligament of malleus

Handle of malleus + tympanic membrane

b

The *stapedius muscle* occupies a bony canal hollowed in the thickness of the posterior wall of the tympanic box. This stapedial canal is situated in front of the upper part of the facial canal. The muscle is inserted at the posterior side of the head of the stapes.

The *tensor tympani muscle* is contained within the bony canal situated at the upper wall of the bony orifice of the auditory tube (Fig. 4.22; 27.b; 38); it is inserted at the upper end of the handle of the malleus.

Imaging

CLINICAL VARIANTS

Whatever the clinical types of affection of the petrous bone, before carrying out special views it is essential to examine the petrous bone by the classic survey views:
- frontal, with projection of the petrous bones into the orbits,
- sagittal (comparative) view,
- Stenvers-Schüller view,
- Worms-Bretton view,
- Meyer view,
- Hirtz view (Fig. 7.27).

These radiographs must be made symmetrically or comparably, save when investigating for fractures. After assessment of the abnormal or suspect zones in these films and of the clinical picture, special and more appropriate views may be used.

These views will depend on the clinical picture:

1) Otosclerosis of the base of the stapes or tympanosclerosis of the external ear:
Radiographic, tomographic and computed tomographic (CT) studies:
- in Guillen transorbital view,
- in Pöschl-Meyer view,
- in symmetric transorbital frontal view.

These display the antro-adito-attical passage, the whole of the auditory ossicles and the space between the base of the stapes and the vestibular (oval) window (Fig. 27–32).

2) Antral or antro-adito-attical cholesteatoma:
Radiographic and computed tomographic (CT) study of the middle ear, the auditory tube from the aditus ad antrum up to the antrum with the tegmen tympani and the tegmen of the external acoustic meatus to display any possible destruction:
- in Stenvers view,
- in Chaussé III view,
- in Guillen view,
- in Pöschl-Meyer view,
- in the 40° opposed transorbital view.

These views are also advisable to complete examination of the superior bulb of the jugular vein when there is a fracture or a tumour of the jugulare glomus that may have affected or destroyed the roof of the jugular foramen.

3) Intracanalicular acoustic neurinoma or fracture of internal ear:
Tomographic study of internal acoustic meatus:
- in Chaussé IV (50° to 60°) view,
- in Stenvers view,
- in Guillen view,
- in Hirtz view,
- in symmetric frontal view (petrous bones projected in orbits).
- in the prestudied inclined sagittal view.

4) Fracture or dislocation of the ossicles or posttraumatic incudomallear or incudostapedial luxation, or ossicular destruction (antro-adito-attical cholesteatoma):
Radiographic study with tomographic and computed tomographic (CT) sections of ossicles:
- in Pöschl-Meyer view,
- in symmetric frontal view with projection of petrous bones into orbits,
- in Guillen comparative view,
- in the prestudied inclined sagittal view.

TECHNIQUE

Tomographic study of the auditory ossicles and middle ear in prestudied inclined sagittal view:
- The subject is in lateral decubitus, with the head in profile resting on the temporoparietal region and making an angle of 30° to 40° in relation to the median sagittal plane;
- the centreing point is situated at the level of the external acoustic meatus to be examined, which is the furthest from the table;
- a tomographic series is to be made starting from the mastoid process, every 5 mm up to the petrous apex; after having identified the best planes for the ossicles several sections are made at 1 mm intervals.

Displayed: This prestudied view allows display of the chain of ossicles along its longest axis and all its bony components (Fig. 29, 30) as well as the stapes and its base opposite the vestibular (oval) window and the lenticular process of the incus. It makes it possible to study: the carotid canal, the jugular foramen, the ostium introitus, the semicircular canals, the vestibule and cochlea, the facial canal and the sulcus of the chorda tympani, the geniculate ganglion, the hiatuses of the greater and lesser petrosal nerves, the antrum, aditus and external acoustic meatus.

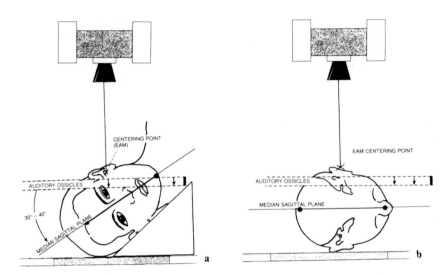

Fig. 28 a, b. Centering diagrams for study of ossicles in prestudied sagittal view

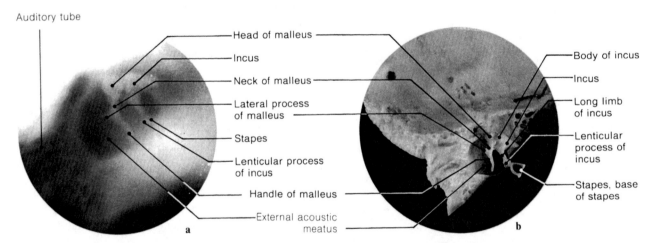

Fig. 29 a Imaging study of anatomic specimen; **b** view of chain of ossicles in dried bone (sagittal views)

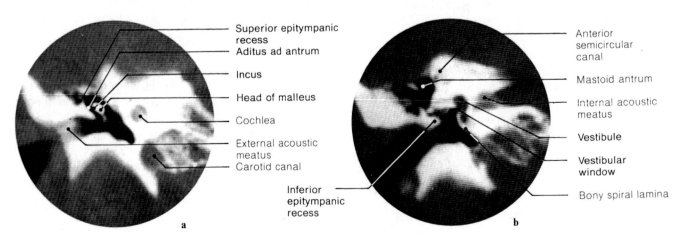

Fig. 30 a, b. CT in varied views; **a** antro-adito-attical passage; **b** vestibular window

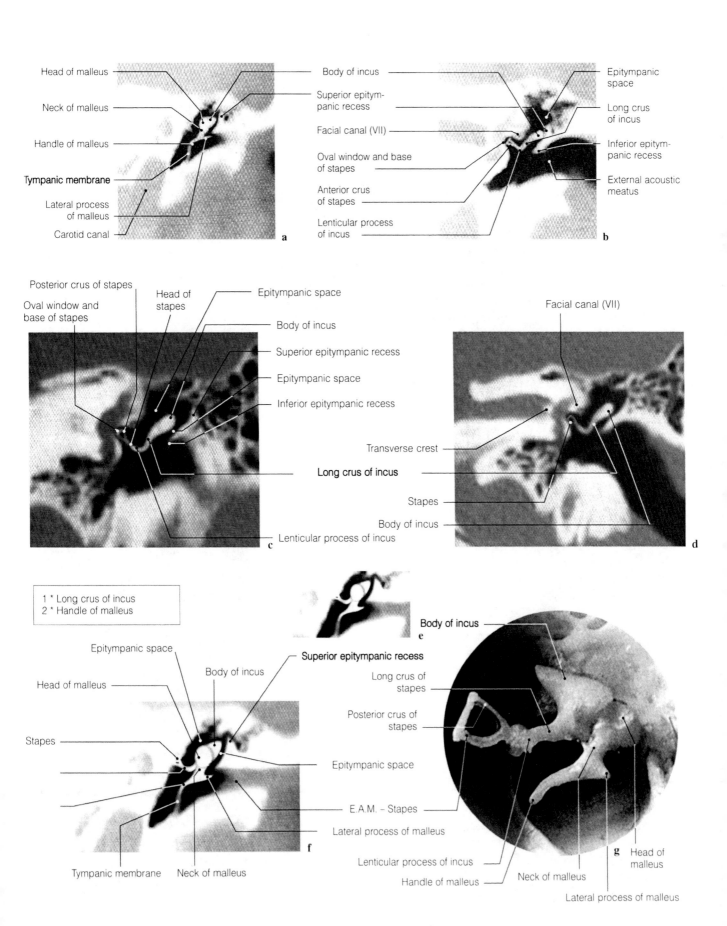

Head of malleus
Neck of malleus
Handle of malleus
Tympanic membrane
Lateral process of malleus
Carotid canal

Body of incus
Superior epitympanic recess
Facial canal (VII)
Oval window and base of stapes
Anterior crus of stapes
Lenticular process of incus

Epitympanic space
Long crus of incus
Inferior epitympanic recess
External acoustic meatus

Posterior crus of stapes
Oval window and base of stapes
Head of stapes
Epitympanic space
Body of incus
Superior epitympanic recess
Epitympanic space
Inferior epitympanic recess
Transverse crest
Long crus of incus
Stapes
Body of incus
Lenticular process of incus

Facial canal (VII)

1 * Long crus of incus
2 * Handle of malleus

Epitympanic space
Head of malleus
Body of incus
Superior epitympanic recess
Stapes
Epitympanic space
E.A.M. – Stapes
Lateral process of malleus
Tympanic membrane   Neck of malleus
Lenticular process of incus
Handle of malleus   Neck of malleus

Body of incus
Long crus of stapes
Posterior crus of stapes
Head of malleus
Lateral process of malleus

Fig. 31 a–g. CT views of the ossicular chain and the cavity of the middle ear along the axis of the footplate of the stapes and the oval window (CT: Dr. J. W. Casselman, A. Z. St Jan, Brugge; Prof. Y. S. Cordoliani; Dr. J. L. Sarrazin, Hôpital du Val-de-Grâce, Paris)

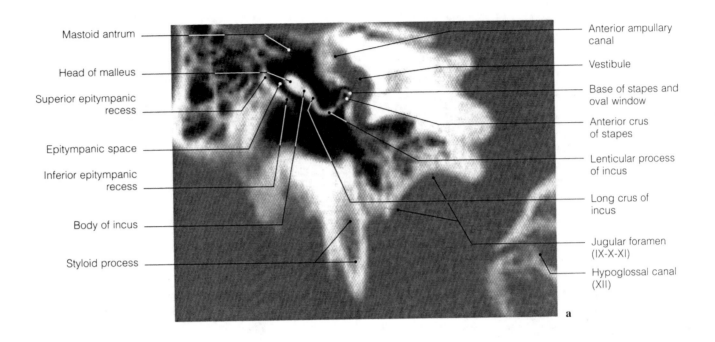

Mastoid antrum

Head of malleus

Superior epitympanic recess

Epitympanic space

Inferior epitympanic recess

Body of incus

Styloid process

Anterior ampullary canal

Vestibule

Base of stapes and oval window

Anterior crus of stapes

Lenticular process of incus

Long crus of incus

Jugular foramen (IX-X-XI)

Hypoglossal canal (XII)

a

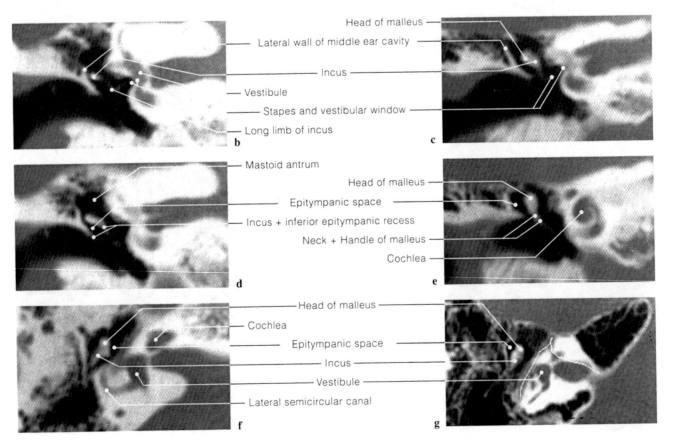

Head of malleus

Lateral wall of middle ear cavity

Incus

Vestibule

Stapes and vestibular window

Long limb of incus

b   c

Mastoid antrum

Head of malleus

Epitympanic space

Incus + inferior epitympanic recess

Neck + Handle of malleus

Cochlea

d   e

Head of malleus

Cochlea

Epitympanic space

Incus

Vestibule

Lateral semicircular canal

f   g

Fig. 32 a–g. CT view of the footplate of the stapes and the oval window (**a**); oblique (**b–e**) and axial (**f, g**) views of the epitympanic space and of the vestibulocochlear pathways

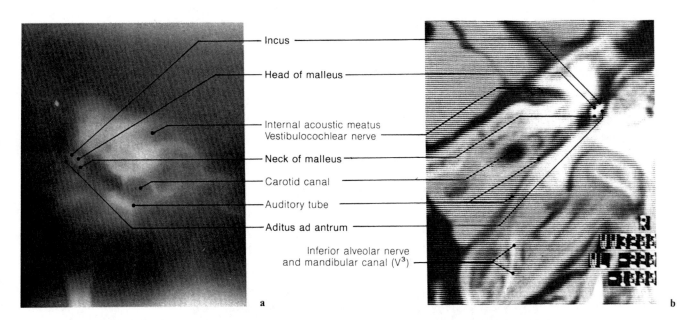

Incus

Head of malleus

Internal acoustic meatus
Vestibulocochlear nerve

Neck of malleus

Carotid canal

Auditory tube

Aditus ad antrum

Inferior alveolar nerve
and mandibular canal (V³)

Fig. 33 a Tomographic Guillen view

Fig. 34 b Computed tomographic (CT)
Worms-Breton view

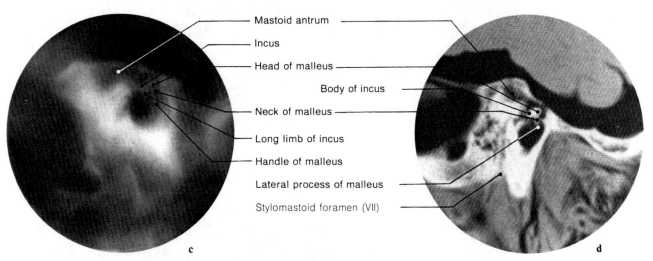

Mastoid antrum

Incus

Head of malleus

Body of incus

Neck of malleus

Long limb of incus

Handle of malleus

Lateral process of malleus

Stylomastoid foramen (VII)

Fig. 35 c, d. Tomography and computed tomography (CT) in prestudied inclined sagittal view

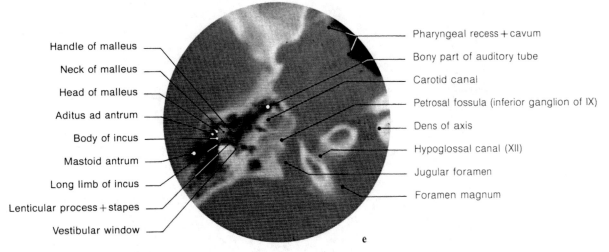

Handle of malleus

Neck of malleus

Head of malleus

Aditus ad antrum

Body of incus

Mastoid antrum

Long limb of incus

Lenticular process + stapes

Vestibular window

Pharyngeal recess + cavum

Bony part of auditory tube

Carotid canal

Petrosal fossula (inferior ganglion of IX)

Dens of axis

Hypoglossal canal (XII)

Jugular foramen

Foramen magnum

Fig. 36 e Computed tomography (CT) in oblique axial view

AUDITORY TUBE
TYMPANIC AND PHARYNGEAL ORIFICES OF THE AUDITORY TUBE, PHARYNGEAL (or ROSENNMÜLLER'S) RECESS
Anatomy, diagrams, and CT views

Anatomy

The auditory tube is the channel for aeration of the middle ear. It connects the tympanic cavity with the rhinopharynx, from which air enters at every swallowing movement to maintain the balance of pressure in the various part of the tympanic cavity.

The auditory tube lies in front of the cavum, aditus and mastoid antrum. It travels obliquely forwards, downwards and inwards.

The auditory tube consists of two parts, bony behind and fibrocartilaginous in front. It is lined by a mucous membrane (Fig. 1; 34; 37–39).

Imaging

CLINICAL FEATURES

When the canal is obstructed, either by a sarcoma or by tumours of cavum, rhinopharynx, jugular glomus or of the pharyngeal recess, this causes displacement of the tympanic membrane towards the interior of the cavity. The equilibrium on both sides of the membrane no longer exists and the stapes becomes embedded in the vestibular window, so producing diminished hearing acuity, vertigo and severe continuous tinnitus. Investigation of the auditory tube then becomes necessary.

Superior petrosal sinus — Internal acoustic meatus

Auditory nerve (VIII) — Labyrinthine part of facialnerve (VII) (I.A.M.)

Superior semicircular canal — First part of facial nerve (VII)

Lateral semicircular canal — Greater superficial petrosal nerve (VII)

Incus — Geniculate ganglion (VII)

Malleus — Auditory tube

External acoustic meatus — Tympanic part of facial nerve (VII), facial canal

Fig. 37 Axial anatomical section at the level of the labyrinth showing the external and internal acoustic meatuses and the vestibulocochlear and facial nerves (anatomical section: Prof. C. Sen, C. S. Chen, K. D. Post, *Microsurgical Anatomy of the Skull Base*, Thieme, 1997)

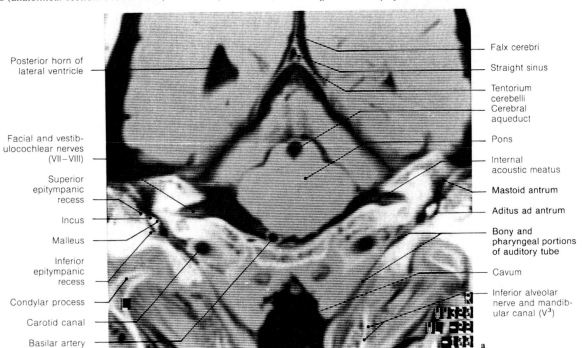

Posterior horn of lateral ventricle — Falx cerebri

Straight sinus

Tentorium cerebelli

Cerebral aqueduct

Facial and vestibulocochlear nerves (VII–VIII) — Pons

Superior epitympanic recess — Internal acoustic meatus

Incus — Mastoid antrum

Malleus — Aditus ad antrum

Inferior epitympanic recess — Bony and pharyngeal portions of auditory tube

Condylar process — Cavum

Carotid canal — Inferior alveolar nerve and mandibular canal (V^3)

Basilar artery

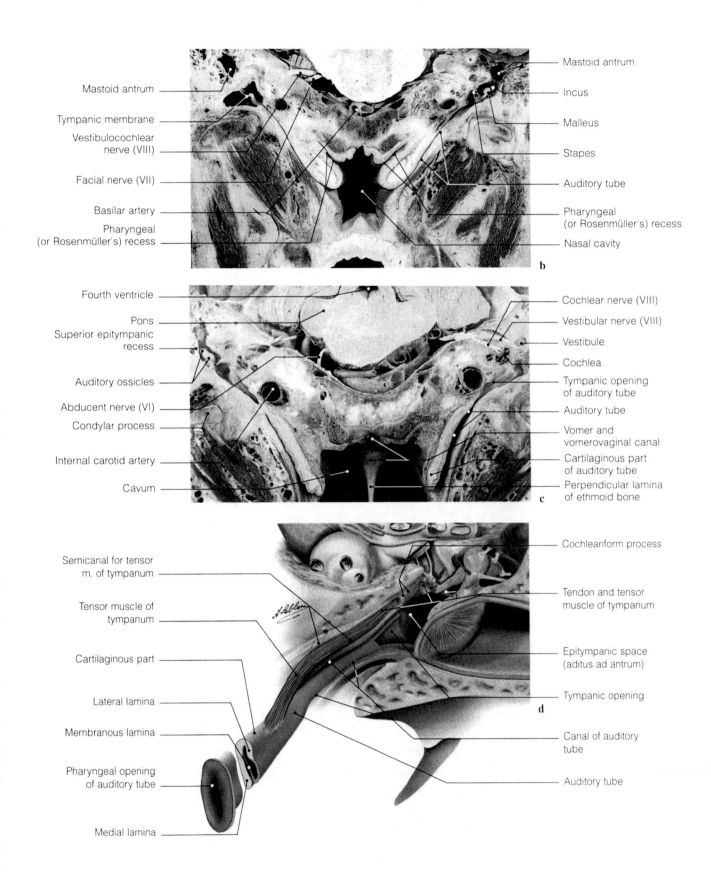

Mastoid antrum

Tympanic membrane

Vestibulocochlear
nerve (VIII)

Facial nerve (VII)

Basilar artery

Pharyngeal
(or Rosenmüller's) recess

Mastoid antrum

Incus

Malleus

Stapes

Auditory tube

Pharyngeal
(or Rosenmüller's) recess

Nasal cavity

b

Fourth ventricle

Pons

Superior epitympanic
recess

Auditory ossicles

Abducent nerve (VI)

Condylar process

Internal carotid artery

Cavum

Cochlear nerve (VIII)

Vestibular nerve (VIII)

Vestibule

Cochlea

Tympanic opening
of auditory tube

Auditory tube

Vomer and
vomerovaginal canal

Cartilaginous part
of auditory tube

Perpendicular lamina
of ethmoid bone

c

Semicanal for tensor
m. of tympanum

Tensor muscle of
tympanum

Cartilaginous part

Lateral lamina

Membranous lamina

Pharyngeal opening
of auditory tube

Medial lamina

Cochleariform process

Tendon and tensor
muscle of tympanum

Epitympanic space
(aditus ad antrum)

Tympanic opening

Canal of auditory
tube

Auditory tube

d

Fig. 38 a–d. CT view (**a**) and anatomical sections (**b, c**) of the auditory tubes and of the epitympanic space; diagrams of the auditory tube from its tympanic ostium to its pharyngeal ostium (**d**)

EXTERNAL EAR
EXTERNAL ACOUSTIC MEATUS
Anatomy, diagrams, and CT views

Anatomy

The external ear is formed of two segments: the auricle and the external acoustic meatus.

The external acoustic meatus is a canal extending from the concha to the tympanic membrane.

The wall of the meatus is cartilaginous and is covered throughout the extent of its internal surface by a skin lining continuous with the skin of the external ear.

The external ear, thanks to its shape and immediate relations, is the receptor organ for sound, which plays a very minimal part in man.

Imaging

CLINICAL FEATURES – INVESTIGATION

In cases of temporal fracture, and if the patient exhibits otorrhagia accompanied by tinnitus with blunted hearing, radiologic examination of the external acoustic meatus is necessary.

It is possible for a temporal fracture to extend into the walls of the external acoustic meatus, involve and displace the inferior epitympanic recess (or wall of the compartment) and produce obstruction of the aditus ad antrum. This fracture may also extend as far as the ossicular chain and produce incudomallear or incudostapedial dislocation; the stapes may be embedded in the vestibular window, producing continuous tinnitus.

Other pathologic conditions of the external acoustic meatus:
– tympanosclerosis,
– an antral and antro-adito-attical cholesteatoma expanding upwards and inferolaterally may, after having destroyed the tegmen tympani, erode the roof of the external acoustic meatus.

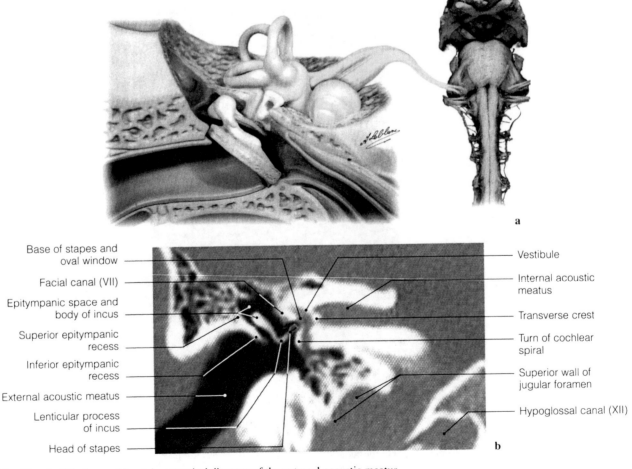

Base of stapes and oval window

Facial canal (VII)

Epitympanic space and body of incus

Superior epitympanic recess

Inferior epitympanic recess

External acoustic meatus

Lenticular process of incus

Head of stapes

Vestibule

Internal acoustic meatus

Transverse crest

Turn of cochlear spiral

Superior wall of jugular foramen

Hypoglossal canal (XII)

Fig. 39 a, b. CT view and frontal anatomical diagram of the external acoustic meatus

References

- Ariyasu L, Galey FR, Hilsinger R Jr, Byl FM (1989) Computer-generated three-dimensional reconstruction of the cochlea. Otolaryngol Head Neck Surg 100:87–91
- Belal A (1979) The effects of vascular occlusion on the human inner ear. J Laryngol Otol 93:955–968
- Brogan M, Chakeres DW (1990) Gd-DTPA-enhanced MR imaging of cochlear schwannoma. AJNR 11:407–408
- Brogan M, Chakeres DW, Schmalbrock P (1991) High-resolution 3DTF MR imaging of the endolymphatic duct and soft tissues of the otic capsule. AJNR 12:1-11
- Byl MF (1977) Seventy-six cases of presumed sudden hearing loss occurring in 1973: prognosis and incidence. Laryngoscope 87:817–825
- Byl MF (1984) Sudden hearing loss: eight years' experience and suggested prognostic table. Laryngoscope 94:647–661
- Casselman JW, Kuhweide R, Ampe W, Meeus L, Steyaert L (1993) Pathology of the membranous labyrinth: comparison of T1–, T2– and Gd-enhanced T1-weighted spin echo imaging and 3 DFT–CISS imaging. AJNR 14:47–57
- Casselman JW, Kuhweide R, Deimling M, Ampe W, Dehaene I Meeus L (1993) Constructive interference in steady state (CISS)–3 DFT MR imaging of the inner ear and cerebello–pontine angle. AJNR 14:59-69
- Casselman JW, Albers FWJ, Major MHJM (1994) MR imaging in patients with Cogan's syndrome. AJNR 15:131–138
- Cole RR, Jahrsdoerfer RA (1988) Sudden hearing loss: an update. Am J Otol 9:211-215
- Daniels DL, Herfkins R, Koehler PR, Millen SJ, Shaffer KA, Williams AL, Haughton VM (1984) Magnetic resonance imaging of the internal auditory canal. Radiology 151:105–108
- Daniels DL, Pech P, Haughton VM (1984) Magnetic resonance imaging of the temporal bone. General Electric Medical Systems Group, Milwaukee
- Daniels DL, Schenck JF, Foster T, Hart H Jr, Millen SJ, Meyer GA, Pech P, Haughton VM (1985) Magnetic resonance imaging of the jugular foramen. AJNR 6:699–703
- Duhamel B, Monod C (1983) Schémas d'anatomie n° 4: Tête et cou (1). Vigot, Paris
- Dulac GL, Claus E, Barrois J (1973) Monographia otoradiologica. Bull Radiogr Agfa Gevaert, août 1973
- Fleury P, François J, Bourdon R (1964) Etude radiotomoanatomique des osselets. Ann Otolaryngol 81:45–52
- Fowler EP (1961) Variations in the temporal bone course of the facial nerve. Laryngoscope 71:937
- Francke JP, Macke A, Clarisse J, Libersa JC, Dobbelaere P (1982) The internal carotid arteries. Anat Clin 3:243-261
- Ge X, Spector G (1981) Labyrinthine segment and geniculate ganglion of facial nerve in fetal and adult human temporal bones. Ann Otol Rhinol Laryngol Supl 85:2
- Galey FR (1984) Initial observations of a human temporal bone with a multi-channel implant. Acta Otolaryngol Suppl (Stockh) 411:38–44
- Guerrier Y (1988) Anatomie chirurgicale de l'os temporal de l'oreille et de la base du crâne, Tome 1 La Simarre, (France)
- Gussen R (1981) Sudden hearing loss associated with cochlear membrane rupture. Arch Otolaryngol 107:598–600
- Hinojasa R, Seligsohn R, Lerner SA (1985) Ganglion cell counts in the cochlea of patients with normal audiograms. Acta Otolaryngol (Stockh) 99:8–13
- Igarashi M, Alford BR, Konishi S, Shaver EF, Guilford FR (1969) Functional and histopathological correlations after microembolism of the peripheral labyrinthine artery in the dog. Laryngoscope 79:603–623
- Jaffe BF (1975) Hypercoagulation and other causes of sudden hearing loss. Otolaryngol Clin North Am 8:395–403
- Johnson DW (1984) Air cisternography of the cerebellopontine angle using high resolution computed tomography. Radiology 151:401–404
- Juster M, Fischgold H (1955) Etude radioanatomique de l'os temporal. Masson, Paris
- Kodros A, Buckingham RA (1957) Anatomy of the descending portion of the facial nerve in the AMA. Arch Otolaryngol 66:735
- Konigsmark BW (1970) Methods for the counting of neurons. In Nauta WJH, Ebbesson SOE (eds) Contemporary Research Methods in Neuroanatomy. Springer Berlin, 315–340
- Korach G, Vignaud J (1977) Manuel de techniques radiographiques du crâne. Masson, Paris
- Kotzias SA, Linthicum FH Jr (1985) Labyrinthine ossification: Differences between two types of ectopic bone. Am J Otol tome: 490–494
- Kudo H, Nori S S (1974) Topography of the facial nerve in the human temporal bone. Acta Anat (Basel) 90:467
- Kumar A, Maudelonde C, Mafee M (1986) Unilateral sensorineural hearing loss: analysis of 200 consecutive cases. Laryngoscope 96:14-18
- Lang J (1981) Facial and vestibulocochlear nerve. Topographic anatomy and variations. In Samii M, Janetta PJ (eds) The Cranial Nerves. Springer, Berlin Heidelberg New York, p 363
- Lang J (1981) Neuroanatomie der N. opticus, trigeminus, facialis, glossopharyngeus, vagus, accessorius und hypo-glossus. Arch Otorhinolaryngol 231:1
- Libersa C (1951) Etude de la vascularisation artérielle des nerfs crâniens et du paquet acoustico-facial. G.Sautai et Fils - Lille
- Linthicum FH Jr, Galey FR (1992) Histologic evaluation of temporal bones with cochlear implants. Ann Otol Rhinol Laryngol tome: 610-613
- Majoor MHJM, Albers FWJ, Casselman JW (1993) The clinical relevance of computed tomography and MR imaging in Cogan's syndrome. Acta Otolaryngol (Stockh) 113:625-631
- Mark AS, Seltzer S, Nelson-Drake J, Chapman JC, Fitzgerald CD, Gulya AJ (1992) Labyrinthine enhancement on gadolinium-enhanced magnetic resonance imaging in sudden deafness and vertigo: correlation with audiologic and electro-nystagmographic studies. Ann Otol Rhinol Laryngol 101:459–464
- Mündnich K, Frey K-W (1959) The Tomogram of the Ear. Das Röntgenschichtbild des Ohres. Thieme, Stuttgart
- Nadol JB Jr, Young YS, Glynn RJ (1989) Survival of spiral ganglion cells in profound sensorineural hearing loss: implications of cochlear implantation. Ann Otol Rhinol Laryngol 98:411–416
- Nager G (1982) The facial canal. Normal anatomy, variations and anomalies. Ann Otol Rhinol Laryngol 91 (Suppl 97):33
- New PFJ, Bachow TB, Wismer GL, Rosen BR, Brady TJ (1985) MR imaging of the acoustic nerves and small acoustic neuromas at 0.6 T: prospective study. AJNR 6:165–170

- Parkin JL, Eddington DK, Orth L, Brackman DE (1985) Speech recognition experience with multi–channel cochlear implant Otolaryngol Head Neck Surg 93:639–645
- Perlmann HB, Kimura R (1957) Experimental obstruction of the venous drainage and arterial supply of the inner ear. Ann Otol Rhinol Laryngol 66:537–546
- Pollak A, Felix H, Schrott A (1987) Methodological aspects of quantitative study of spiral ganglion cells. Acta Otolaryngol Suppl. (Stockh) 436:37–42
- Proctor B, Nager G (1982) The facial canal. Normal anatomy, variations and anomalies. Ann Otol Rhinol Laryngol 91 (Suppl 97):33
- Rabischong P, Vignaud J, Paleirac R, Lamoth AP (1975) Tomographie et anatomie de l'oreille. Lamoth, Amsterdam
- Sanna M, Saleh E, Russo A, Taibah A (1995) Atlas of Temporal Bone and Lateral Skull Base Surgery. Thieme, Stuttgart
- Schuknecht HF, Donovan ED (1986) The pathology of idiopathic sudden sensorineural hearing loss. Arch Otorhinolaryngol 243:1–15
- Seltzer S, Mark AS (1991) Contrast enhancement of the labyrinth on MR scans in patients with sudden hearing loss and vertigo: evidence of labyrinthine disease. AJNR 12:13–16
- Sen C, Chen CS, Post KD (1997) Microsurgical anatomy of the skull base. Thieme, Stuttgart
- Simmons FB (1979) The double-membrane breaks syndrome in sudden hearing loss. Laryngoscope 89:59–66
- Suga F, Preston J, Snow JB Jr (1970) Experimental microembolization of cochlear vessels. Arch Otolaryngol 92:603–623
- Tanioka H, Shirakawa T, Machida T, Sasaki Y (1991) Three dimensional reconstructed MR imaging of the inner ear. Radiology 178:141–144
- Van Dishoeck HAE, Bierman TA (1957) Sudden perspective deafness and viral infection. Ann Otol Rhinol Laryngol 66:959–969
- Valavanis A, Schubiger O (1983) High-resolution CT of the normal and abnormal fallopian canal. Am J Neuroradiol 4:748
- Valvassori GE (1976) Radiography of the facial nerve canal. In Fisch U (ed) Proceedings of the 3rd Symposium on Facial Nerve Surgery, August 1976, Zurich, Switzerland, p 174 (abstract)
- Valvassori GE, Potter DG, Hanafee WN, Carter BL, Buckingham RA (1982) Radiology of the Ear, Nose and Throat. Saunders, Philadelphia
- Vignaud J, Korach G (1969) Exploration radiologique du rocher normal. Feuill Electroradiol 51:52
- Vignaud J, Sultan A, Leriche H (1969) Dislocations traumatiques de la chaîne des osselets. J Radiol Electro 50:803–806
- Vignaud J, Jardin C, Rosen L (1986) The Ear–Diagnostic Imaging. Masson, New York
- Wadin K, Wilbrand H (1987) The labyrinthine portion of facial canal. A comparative radioanatomic investigation. Acta Radiol 28:17–32
- Weill F (1975) Eléments programmés de radiologie oto-rhino-sto-matologique. Masson, Paris
- Wilbrand HF (1974) Multidirectional Tomography of Minor Detail in the Temporal Bone. Acta Universitatis Upsaliensis, Upsala

Index

- Abducent nerve   9, 10, 11, 14, 40–49
- Accessory nerve   9
- Aditus ad antrum   35, 44–49
- Alveolar nerve, inferior   47
- Ampulla of posterior semicircular duct   22, 23
- Ampullae of ampullary crest   23, 24, 30
- Ampullary canal, anterior   43
- Ampullary crest   22–24, 30
- Ampullary nerve, anterior   7, 15–17, 22–24, 31, 35
- Ampullary nerve, lateral   7, 15-17, 22–24, 31, 35
- Ampullary nerve, posterior   7, 15-17, 22–24, 31, 35
- Ampullary nerve, posterior, in foramen singulare   6, 35, 39
- Ampullary sulcus   22, 23
- Andersch's ganglion (IX)   29
- Anterior genu of internal carotid artery   40
- Aperture of cochlear aqueduct   26–31
- Aperture of vestibular aqueduct   26–31
- Apparent origin of vestibulocochlear nerve (bulbopontine sulcus)   6–8
- Aqueduct of mesencephalon   48
- Arachnoid   29
- Area of facial nerve   15
- Auditory tube   3, 7, 35, 42, 47–49

- Basilar artery   9-12, 14, 15, 34, 36–40
- Basilar membrane of cochlear duct   19–21
- Bony crest   23
- Bony labyrinth   15, 26–32

- Capillary network in cochlear aqueduct   30, 31
- Capillary network in vestibular aqueduct   30, 31
- Capillary network in helicotrema   29, 31
- Carotid artery, internal   2, 40, 49
- Carotid artery, intracavernous   47, 49
- Carotid canal   45, 47–49
- Cartilaginous part of auditory tube   49
- Cavity of geniculate ganglion   13, 48
- Cells of Claudius   19–21, 23
- Cells of Deiters   19–21, 23
- Cells of Hensen   18–22, 24
- Cerebellar artery, anterior inferior   9–12, 32–41
- Cerebellar artery, superior   34, 36
- Cerebellar peduncle, inferior   5
- Cerebellopontine angle   4, 10, 11
- Cerebellum   10, 18–20
- Cerebral artery, posterior   9, 34
- Cerebral peduncle   5
- Clivus   36
- Cochlea   7, 14, 16–22
- Cochlear aqueduct   26, 27, 30, 31
- Cochlear area   15
- Cochlear artery   36–39
- Cochlear artery, common   34, 39

- Cochlear artery, posterior   32–34, 39
- Cochlear branch of the vestibulocochlear artery   33, 39
- Cochlear canaliculus   29–31
- Cochlear cavities   17–21
- Cochlear duct   30, 31
- Cochlear ganglion (Spiral ganglion of cochlea)   15, 17–25
- Cochlear modiolus   25
- Cochlear nerve   4, 9-12, 36–39
- Cochlear nuclei   4, 5
- Cochlear nucleus, anterior   4, 5
- Cochlear nucleus, posterior   4, 5
- Cochleariform process   3, 41, 49
- Columella   29, 31
- Common semicircular bony canal   30
- Communicating artery, posterior   34
- Condylar process of mandible   49
- Cupula of cochlea   21, 22, 25

- Endolymphatic duct   29, 30
- Endolymphatic sac   29, 30
- Epitympanic recess, inferior   41–55
- Epitympanic recess, superior   41–55
- Epitympanic space   22, 24, 30, 31
- External acoustic meatus   41–49
- External ear   4, 7, 35, 48
- External oral cavity   45, 46

- Facial canal   45, 48
- Facial nerve   7, 11, 15, 35–38, 48, 49
- Fibres for the superior vestibular nerve   23, 24
- Floor of the fourth ventricle   4, 5
- Fossula of cochlear window   29–31
- Fourth ventricle   10–12, 36, 38
- Frenulum of superior medullary velum   4, 5

- Geniculate ganglion   48
- Glossopharyngeal nerve   9, 35

- Hair cells in the ampullary crest   22–24, 30–32
- Hair cells in the spiral organ of Corti, inner   18–20
- Hair cells in the spiral organ of Corti, outer   18–20
- Hypoglossal canal   9, 47, 50
- Hypoglossal nerve   9, 35

- Incudomallear articulation   41–50
- Incudostapedial articulation   41–50
- Incus   41-50
- Inferior colliculus   5
- Infrastrial capillaries   20
- Inner tunnel (Corti's tunnel)   19, 20, 24
- Interdental cells   18–22, 24
- Intermediate nerve   3, 7, 10, 11
- Internal acoustic meatus   44–49

• Internal ear 4, 7, 35, 48

• Jugular foramen 9, 13, 29, 31, 44, 46, 50

• Labyrinthine artery 36–39
• Labyrinthine vasculature 32–40
• Lamina of cartilage of auditory tube, lateral 49
• Lamina of cartilage of auditory tube, medial 49
• Lenticular process of incus 35, 41–45, 49, 50
• Ligament of stapes, lateral 35, 41
• Ligament of stapes, superior 35, 41
• Ligament of incus, posterior 35, 41
• Ligament of incus, superior 35, 41

• Macula saccule 30
• Macula utricle 30
• Malleus 3, 7, 19, 41–47
• Malleus, body of 41–46
• Malleus, head of 3, 7, 19, 35, 41–47
• Malleus muscle 7, 35
• Malleus, neck of 41–47
• Mandibular canal 48
• Manubrium of malleus 35, 41–45
• Mastoid antrum (tympanic antrum) 7, 35, 41–47
• Membranous labyrinth 16, 26–32
• Membranous lamina of auditory tube 3, 49
• Membranous wall of utricle 7, 16, 27
• Middle ear 35, 41–50
• Modiolus 25

• Neuroepithelium 6, 22, 23

• Occipital foramen 9, 12
• Oculomotor nerve 6, 9, 34
• Ophthalmic nerve 36
• Optic nerve 6, 34, 40
• Ossicles 41–50
• Ossicular chain 41–50
• Otolythic membrane 22
• Oval window (fenestra vestibuli) 7, 32, 35, 41, 45

• Perilymphatic duct 29, 31
• Perilymphatic space 22, 26, 27, 30, 31, 35
• Perpendicular lamina of ethmoid bone 3, 49
• Petrosal fossa 14, 29–31, 44
• Petrosal sinus, inferior 9
• Petrosquamous fissure 31
• Pharyngeal recess 49
• Pillar cells, inner (Corti's rods) 14, 19
• Pillar cells, outer (Corti's rods) 14, 19
• Pituitary gland 40
• Pons 3-12, 34–41
• Pons, inferior 3-12, 34–41

• Posterior horn of lateral ventricle 4, 5
• Process of malleus, anterior 3, 7, 35, 41–47
• Process of malleus, lateral 7, 19, 35, 41–47

• Radial artery, external 21, 39
• Radial artery, internal 21, 39
• Radicular fibres for the hypoglossal nerve 9, 40
• Real origins of the vestibular and cochlear nerves (vestibular and cochlear nuclei) 4, 5
• Recurrent artery 34, 37–39

• Saccular area 15
• Saccular canal 26, 27, 30, 31
• Saccular nerve 6, 15, 17, 22, 35, 39
• Saccule 6, 15, 17, 22, 35, 39
• Scala tympani 14–22, 24, 25, 32, 35, 39
• Scala vestibuli 14–22, 24, 25, 32, 35, 39
• Secondary spiral lamina 25
• Secondary tympanic membrane (membrane of round window) 31
• Semicanal of tensor tympani muscle 46
• Semicircular canals 3, 16–18, 22, 25–30, 41, 43
• Semi–oval fossula 26
• Spiral canal of cochlea 25
• Spiral canal of modiolus 25
• Spiral lamina of cochlea 20, 21, 25
• Spiral limbus 21, 39
• Spiral organ of Corti 7, 14, 15–25
• Spiral vein, anterior 21, 39
• Spiral vein, posterior 21, 39
• Stapedius muscle 7, 35
• Stapes 41–46, 49, 50
• Stapes, base of 41–46, 49, 50
• Stapes, foot of 41–46
• Stapes, head of 41–46, 49, 50
• Stereocilia 19, 22
• Straight sinus 13
• Stria vascularis of cochlear duct 21, 39
• Styloid process 3, 49
• Stylomastoid artery 32
• Subarachnoid 29
• Subarcuate artery 34, 37–39
• Subvestibular space 26–31

• Tectorial membrane 6, 14–22
• Tectorial membrane of cochlear duct (Corti's membrane) 14–22
• Tendon of tensor tympani muscle 3, 49
• Tensor tympani muscle 3, 7, 35, 49
• Transverse crest of internal auditory meatus (crista transversa) 15
• Transverse foramen of vertebral artery 15
• Transverse sinus 13

- Trigeminal nerve   1, 6, 9
- Trochlear nerve   6, 9
- Tympanic artery, inferior   32
- Tympanic canaliculus   29, 31
- Tympanic cavity   41, 42
- Tympanic membrane   3, 7, 35, 41, 42, 45–49

- Utricle   7, 15, 16, 22, 27–31
- Utricular canal   26, 27, 30, 31
- Utricular nerve   7, 17, 35
- Utricular segment   22, 27, 29–31
- Utriculosaccular duct   27, 30, 31

- Vagus nerve   5, 6
- Vein of cochlear aqueduct   29
- Vein of vestibular aqueduct   29
- Vein of endolymphatic sac   28
- Vein of modiolus, common   21, 39
- Vein of round window   33
- Vein of semicircular canals   28, 29
- Vertebral artery   15, 35, 40
- Vestibular aqueduct   26, 27, 30, 31
- Vestibular area, inferior   15
- Vestibular area, superior   15
- Vestibular artery, anterior   21, 39

- Vestibular artery, superior   21, 39
- Vestibular branch of the vestibulocochlear artery   32, 33
- Vestibular cecum of cochlear duct   30, 31
- Vestibular ganglion   17, 18, 22
- Vestibular nerve   4, 9–12, 36–39
- Vestibular nerve, inferior   8–12, 37, 38
- Vestibular nerve, superior   8–12, 37, 38
- Vestibular nuclei   4, 5
- Vestibular nucleus, inferior   4, 5
- Vestibular nucleus, lateral   4, 5
- Vestibular nucleus, medial   4, 5
- Vestibular nucleus, superior (Bekhterev's nucleus)   4, 5
- Vestibular vein, anterior   29
- Vestibular vein, posterior   29
- Vestibular wall of cochlear duct (Reissner's membrane)   6, 14–22
- Vestibule   19, 22, 32, 39, 46, 50
- Vestibulocochlear artery   17, 33
- Vestibulocochlear branch   33, 35, 39
- Vestibulocochlear nerve   8-12, 37, 38
- Vomer   49

- Wall of auditory tube   7, 35, 47–49
- Wall of epitympanic recess   7, 35, 47–49

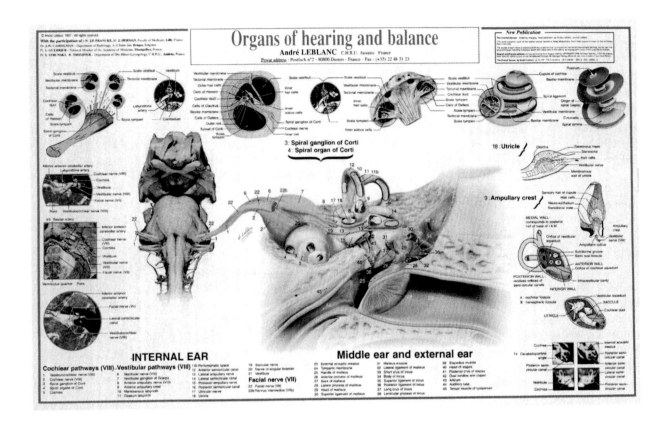

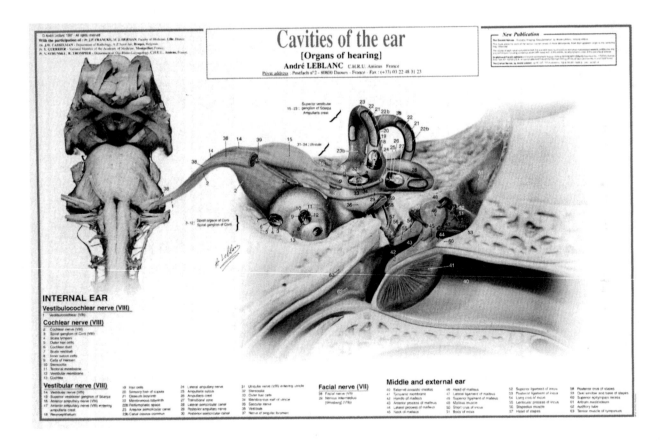

If you wish to place an order for posters, please contact the author at the following address:
André Leblanc, Boîte Postale n°2, 80800 Daours, France - Fax: (33) 3 22 48 31 23

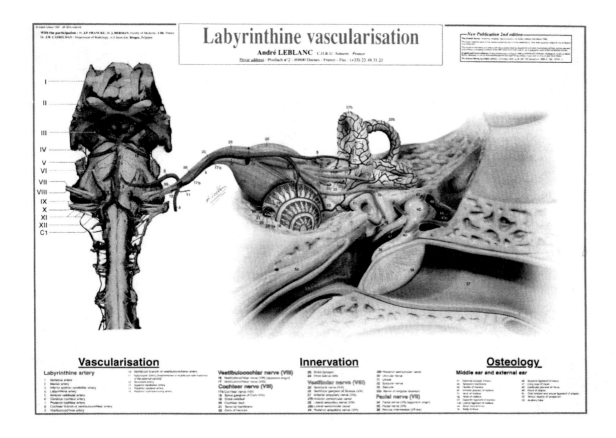

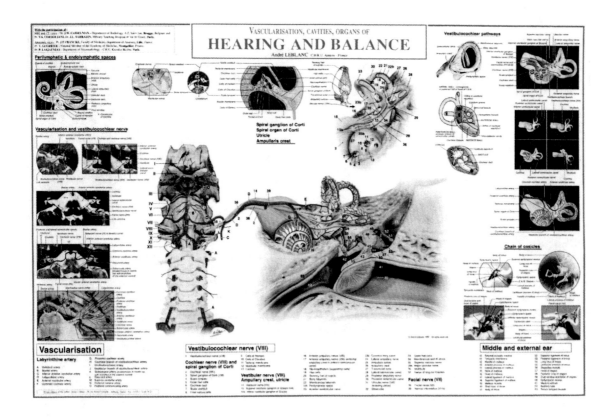

If you wish to place an order for posters, please contact the author at the following address:
André Leblanc, Boîte Postale n°2, 80800 Daours, France - Fax: (33) 3 22 48 31 23

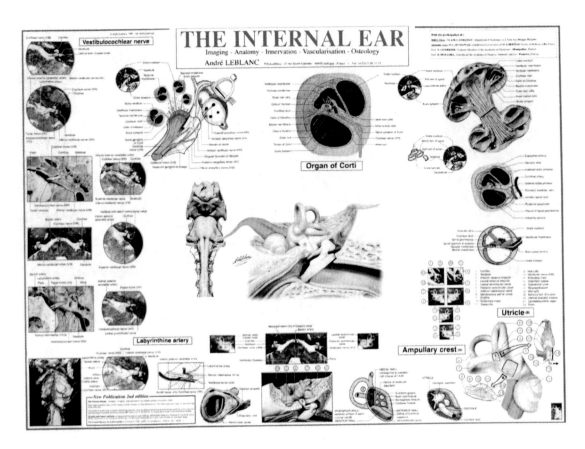

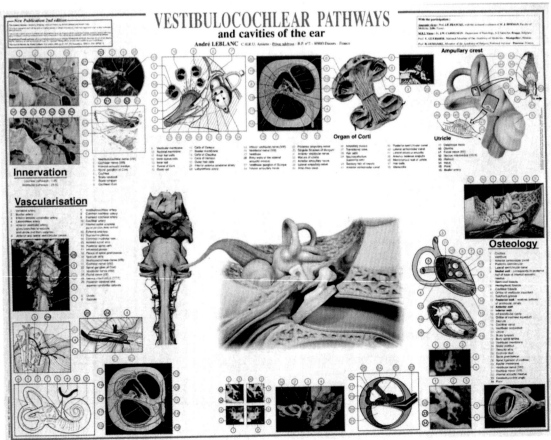

If you wish to place an order for posters, please contact the author at the following address:
André Leblanc, Boîte Postale n°2, 80800 Daours, France - Fax: (33) 3 22 48 31 23

Imprimé par JOUVE – 18, rue Saint-Denis, 75001 PARIS
N° 260166R – Dépôt légal : septembre 1998